Fondazione Renato Giunti

—

GALILEO'S TELESCOPE

Galileo's Telescope

The Instrument that Changed the World

Edited by
Giorgio Strano

GIUNTI

Galileo's Telescope
The Instrument that Changed the World

4 March– 31 December 2008

Galileo's Telescope is part of the series of events celebrating the 400th anniversary of Galileo's celestial discoveries

Promoting institutions

Istituto e Museo di Storia della Scienza

Ministero dell'Istruzione dell'Università e Ricerca

Regione Toscana

Ente Cassa di Risparmio di Firenze

Under the patronship of

Provincia di Firenze

Comitato Nazionale per le Celebrazioni Galileiane 2009-2010

With the collaboration of

OFFICINE PANERAI

Opera Laboratori Fiorentini SpA

Exhibition scientific project
Istituto e Museo di Storia della Scienza

With the collaboration of
Istituto Nazionale di Fisica Nucleare, Florence
Istituto Nazionale di Ottica Applicata
Osservatorio Astrofisico di Arcetri
Stazione Sperimentale del Vetro, Murano

Scientific coordination and curatorship
Giorgio Strano, Istituto e Museo di Storia della Scienza

Scientific committee
Filippo Camerota, Istituto e Museo di Storia della Scienza
Paolo Del Santo, Regione Toscana
Sven Dupré, Universiteit Ghent
Paolo Galluzzi, Istituto e Museo di Storia della Scienza e Università di Firenze
Pier Andrea Mandò, Istituto Nazionale di Fisica Nucleare e Università di Firenze
Giuseppe Molesini, Istituto Nazionale di Ottica Applicata
Francesco Palla, Osservatorio Astrofisico di Arcetri
Albert Van Helden, Universiteit Utrecht
Marco Verità, Stazione Sperimentale del Vetro, Murano

Organisation
Istituto e Museo di Storia della Scienza:
Laura Manetti

Exhibition set up
Opera Laboratori Fiorentini S.p.A., Florence

Graphic design
RovaiWeber design, Florence

Replicas and working models
Concept and scientific planning
Giorgio Strano, Paolo Del Santo,
Giuseppe Molesini, Paolo Stefanini
Feasible project and modeling
Opera Laboratori Fiorentini S.p.A., Florence
Jim & Rhoda Morris, Wakefield, USA

Multimedia
Concept and iconographic research
Giorgio Strano, Paolo Del Santo
Production
Istituto e Museo di Storia della Scienza
multimedia lab: Daniela Vespoli, with
Jacopo Tonini, Monica Tassi, Riccardo Braga

Web site
Istituto e Museo di Storia della Scienza:
Marco Berni, with Fabrizio Butini,
Leonardo Curioni, Elena Fani, Damiana
Luzzi, Roberta Massaini, Iolanda Rolfo

Iconographic coordination
Istituto e Museo di Storia della Scienza:
Franca Principe, with Sabina Bernacchini,
Susanna Cimmino

Press office
Catola & Partners, Florence

English translations
Catherine Frost, with Karen Tomashavsky

Italian translations
Giorgio Strano

We thank for their collaboration
Biblioteca Marucelliana, Florence
Biblioteca Nazionale Centrale, Florence
Biblioteca Riccardiana, Florence
Dipartimento di Elettronica e Informazione,
Politecnico di Milano
Istituto Nazionale di Fisica Nucleare, Milan
Roberto Alberti, Novella Grassi,
Antonio Longoni, Alessandro Migliori

Photografic credits
Accademia dei Lincei, Rome
Algemeen Rijksarchie, The Hague
Biblioteca Comunale degli Intronati, Siena
Biblioteca Nazionale Centrale, Florence
Biblioteca Riccardiana, Florence
Bibliothèque Inguimbertine, Carpentras
Istituto e Museo di Storia della Scienza,
Florence
Soprintendenza Archeologica per la Toscana,
Florence
Universiteitsbibliotheek, Leiden

Contributors to the catalogue
Paolo Del Santo (P.D.S.)
Sven Dupré (S.D.)
Simone Esposito (S.E.)
Pier Andrea Mandò (P.A.M.)
Luca Mercatelli (L.M.)
Giuseppe Molesini (G.M.)
Jim & Rhoda Morris (J.& R.M.)
Giorgio Strano (G.S.)
Albert Van Helden (A.V.H.)
Maurizio Vannoni (M.Va.)
Marco Verità (M.Ve.)

The exhibition on the Internet
http://brunelleschi.imss.fi.it/telescopiogalileo

Catalogue
Giunti Editore

Editorial Manager
Claudio Pescio

Editor
Dario Dondi

Translation
Catherine Frost, with Karen Tomashavsky

Graphic Design and Layout
RovaiWeber Design

Technical Coordinator
Alessio Conticini

Photolithography
Fotolito Toscana, Florence

via Bolognese 165 - 50139 Florence - Italy
via Dante 4 - 20121 Milan - Italy
www.giunti.it

First edition: March 2008

Reprint	Year
5 4 3 2 1 0	2011 2010 2009 2008

Printed by Giunti Industrie Grafiche S.p.A. – Stabilimento di Prato

Index

Presentation

Paolo Galluzzi
Director
Istituto e Museo di Storia della Scienza

The exhibition *Galileo's Telescope* marks the opening of a series of ambitious events planned by the Istituto e Museo di Storia della Scienza in Florence to celebrate in innovative manner the fourth centennial of Galileo's astronomical utilisation of the telescope and the stunning discoveries he made with it. In the exhibition's subtitle, the telescope is rightly called "The instrument that changed the world". This is by no means exaggerated, when we consider that the new realms observed in the sky by Galileo produced a real revolution, which not only helped to demolish the traditional images of the universe but also had traumatic repercussions on the philosophical, religious and anthropological levels. Casting doubt on the centrality and immobility of the Earth led in fact to formulating new and dramatic questions as to the origins of the world and the sense and purpose of human life, suggesting answers less satisfying and comforting than those of the ancient and medieval philosophical traditions. The wounds provoked by that sudden earthquake left deep marks on the process of formation of the modern culture, radically undermining convictions that were ancient and deep-rooted. Above all, intense conflict was triggered between science and religion, conflict that in different forms and degrees of intensity has continued down to our own time.

All of this upheaval and the enormous load of epochal consequences on the cognitive level, in the perception of the meaning of the world and the destiny of mankind, was the consequence of the insight of a man of genius who transformed a curious object, sold as a toy in late 1608 and early 1609 in the narrow streets of Venice and many other European cities, into a refined instrument. Strange as it seems, the idea of pointing the new instrument at the sky had not crossed the minds of the Dutch makers who had developed the first rudimentary models of the telescope. Thanks to Galileo's brilliant inspiration, in autumn 1609 a new world began to unfold before his astonished eyes. It was a world that many of his colleagues refused to even look at, while others, many more, went to absurd lengths to demonstrate that the celestial objects brought down to Earth by the optical tube did not change what was already known by one iota. For them, no effort need be made to revise traditional knowledge on the book of nature. But the discoveries made by Galileo with the telescope were to change the world much more drastically than the geographical discoveries of Columbus, Vespucci and the great navigators who followed in their wake.

It is thus not merely opportune but necessary to inaugurate a program of celebrations of Galileo's astronomical discoveries with an exhibition centred on a thoughtful, well-planned presentation of the Galilean telescope, delineating its physiology, illustrating its anatomy, inserting it within its historic context and rigorously analysing its structure and performance through the most advanced systems of scientific investigation available today.

Confronted in a strictly interdisciplinary spirit, the history of the astronomical utilisation of the telescope becomes a fascinating adventure, attracting the interest of a vast public in Italy and abroad, as shown by the innumerable requests from important partners to present the exhibition at other venues as well.

The preparation of an organic exhibition based on integrated historical and scientific analyses and designed to involve the visitor as protagonist, allowing him to relive the emotion of Galileo's astronomical discoveries, would have been unthinkable without the collaboration of a broad range of experts, all of the highest level, such as are only rarely found working together on the same project.

I would therefore like to begin a long series of acknowledgements with thanks to my colleagues Enzo Iarocci (President of the Istituto Nazionale di Fisica Nucleare), Francesco Palla (Director of the Osservatorio Astrofisico in Arcetri), Paolo De Natale (Director of the Istituto Nazionale di Ottica Applicata), and Antonio Tucci (Director of the Stazione Sperimentale del Vetro in Murano), for having promptly accepted my invitation to involve their prestigious research institutes in our project of "global" exploration of Galileo's telescope, the only two surviving examples of which (in addition to the objective lens of the "excellent" instrument he used to discover the satellites of Jupiter) constitute the most precious heritage of the Istituto e Museo di Storia della Scienza. The contributions of the scholars and technical staff of the above-mentioned institutions have been crucially important. The exhibition and its catalogue reveal the exceptionally profound reading that results from the programmatic integration of advanced scientific investigations with in-depth historical/philological research. But the conclusions presented here, which have been drawn only after a few months of work, are provisional. Ongoing research will undoubtedly lead to further acquisitions, thus intensifying public interest in Galileo's telescope by calling it to the attention of the whole world as a "cultural asset" of universal value.

I would like to express special thanks to Giorgio Strano, our exceptionally able Curator, who has masterfully coordinated this complex operation, assuming responsibility for the exhibition and the catalogue, with the expert assistance of Italian and foreign collaborators of the highest level. To those colleagues who have made available their expertise, contributing to the research for and preparation of the exhibition and its catalogue, go my deep appreciation as well as admiration. Laura Manetti has followed the complex organisation of the event with her customary precision and efficiency.

The Istituto e Museo di Storia della Scienza has been able to undertake this demanding project and bring it to conclusion thanks to the indispensable support of numerous public and private bodies. Research on the project began with an official agreement stipulated with the Ministero dell'Università e della Ricerca. Through a new agreement, recently signed, the exhibition on Galileo's telescope will travel to Beijing. The Ente Cassa di Risparmio di Firenze has firmly supported this project, within the context of its vitally important backing of initiatives for promoting the Istituto e Museo di Storia della Scienza in the last two decades. The Provincia di Firenze and the Regione Toscana have encouraged the line of research on the telescope, supporting in particular the development of incisive educational activities addressed to youth and the scholastic world. A private sponsor of great international prestige, Officine Panerai, which boasts unforgotten Florentine roots and shares the Galilean urge toward innovation, has shown deeply appreciated interest in our exhibition project, offering support also as regards travelling plans for our exhibition. Lastly, I am grateful to Opera Laboratori Fiorentini for having designed, in very brief time and with generous deployment of means, a fascinating exhibition itinerary.

Galileo's Telescope has been organised under the aegis and with the sponsorship of the Committee for the Celebrations of the Fourth Centennial of Galileo's astronomical discoveries. Under this prestigious label – along with many other major events now being prepared in Pisa, Padua, Rome, and numerous other places in Italy and the rest of the world – other highly interesting initiatives are being planned by the Istituto e Museo di Storia della Scienza for this epochal occasion. In March 2009 the great exhibition *The Universe of Galileo. Images of the Cosmos from Antiquity to the Scientific Revolution* will be inaugurated in Palazzo Strozzi, promoted by the Palazzo Strozzi Foundation, the Ente Cassa di Risparmio di Firenze and the Region of Tuscan, and entrusted to the scientific responsibility of our institute. In autumn 2009 the permanent exhibition spaces of our institution will reopen, after radical restructuring of the building, presenting a totally new look and assuming the new name of “Museo Galileo. Istituto di Storia della Scienza”.

The programme I have outlined represents our contribution to the solemn anniversary of 2009-2010. We hope to be able to honour the achievements of this great mind in an adequate manner, by resurrecting the spirit of industrious commitment and programmatic innovation that was the salient feature of Galileo’s inimitable personality.

PRESENTATION

Edoardo Speranza
President
Ente Cassa di Risparmio di Firenze

It is not surprising that human ingenuity can produce inventions whose importance and consequences for technical and scientific progress go initially unrecognised. It is instead surprising and fascinating that a few elect minds can instinctively recognise the potential value of instruments that others have fabricated almost by chance, or for much less ambitious ends.

Genius resides instead in the creative force of imagination that becomes reality, a theoretical intuition that is transformed into a revolution, overthrowing rules and beliefs that had been held definitive and immutable, the courage of reason applied to a magic tube that expands the boundaries of the universe and broadens the horizons of knowledge: Galileo in search of the truth as opposed to any plausible accepted certainty – this is the true dimension of the scientist, and for this reason he belongs to all time.

We are glad that the calendar of scientific celebrations crosses our path today with one of the most prestigious events for Florence and Tuscany as a whole: the fourth centennial of Galileo's astronomical discoveries of 2009, for which the exhibition *Galileo's Telescope* represents both preview and prologue, the stimulating premise for a unique opportunity that will be admirably promoted by the Istituto e Museo di Storia della Scienza.

The Ente Cassa di Risparmio di Firenze shares in the projects of this worthy institution which, through its educational attention to scientific knowledge, engages the planet in dialogue and, never content to rest on the laurels of goals attained, ceaselessly renews itself, as it is now about to do with the redesigning of its permanent collections exhibition, showing a vital dynamism more often found in a laboratory than in the hushed atmosphere of a museum.

The Galilean celebrations thus offer still another occasion and, in this case the best, to renew the invitation, natural for us as Florentines, to participate in a history of scientific evolution that is the common heritage of mankind and the incentive that arouses curiosity and ardent enthusiasm for the phenomena that move the world.

SEXTANS ASTRONOMICUS TRIGONICUS
PRO DISTANTIIS RIMANDIS.

Introduction:
The Two ways of Astronomical Observation

Giorgio Strano

In the 16th century, traditional astronomy was two-fold. On the one hand, 'cosmology' was a matter for the philosopher, dealing with the 'real' structure of the Universe. According to Aristotle (384-322 B.C.E.), the Universe was believed to be formed of a system of concentric crystalline spheres, carrying all the planets uniformly around the central immobile Earth. On the other hand, 'mathematical astronomy' was a matter for the mathematician, dealing with the definition of geometrical models useful in making calculations to predict the position of the planets along the Zodiac with the highest accuracy. Part of mathematical astronomy, the so-called 'observational astronomy' used reliable graduated instruments to provide positional data **(Fig. 0.1)**.

Usually, cosmology and mathematical astronomy did not interact. The mathematician had to forecast planetary positions independently from the actual structure of the Universe. The philosopher had to determine the physical structure of the Universe independently from forecasting planetary positions. Only a few successful attempts were made to overcome this dichotomy. In *On the Revolutions of the Heavenly Spheres* (1543) Nicolaus Copernicus (1473-1543) delineated the heliocentric system of the Universe by re-arranging and correcting the geometrical planetary models of traditional mathematical astronomy. In *On the Most Recent Phenomena of the Ethereal World* (1588) Tycho Brahe (1546-1601) demonstrated that the Aristotelian structure of the Universe was untenable, on the bases of careful determination of the distance of novae and comets from the Earth.

Galileo Galilei (1564-1642) also made early attempts to merge the two branches of traditional astronomy. In 1597, he was convinced of the correctness of the heliocentric structure of the Universe, as he confessed in a letter to Johann Kepler (1571-1630). However, Galileo also tried to prove the validity of the Copernican theory on the bases of angular measurements. In 1604, a nova appeared in the constellation Ophiucus. This phenomenon gave Galileo the idea to determine the Earth's

Fig. 0.1 – Tycho's astronomical sextant (W. J. Blaeu, *Le Grand Atlas, ou Cosmographie ...*, 1667).

movement around the Sun by minimal annual changes in the apparent position of the nova. However, Galileo was not a skilled observer in the established practical astronomy, and preferred to examine – unsuccessfully – the data gathered by others on the nova that appeared in 1572 in the constellation Cassiopeia. This mediation between cosmology and mathematical astronomy singles out Galileo's early tendency to consider himself a 'natural philosopher': an individual able to investigate the real structure of the Universe by using both philosophy and mathematics.

Galileo's fortuitous learning of the optical device, that, in his hands, became the astronomical telescope **(Fig. 0.2)**, enabled him to overcome the many difficulties of traditional astronomy in an unexpected manner. Instead of permitting reliable angular measurements of star and planetary positions, as in usual observations accomplished up to the time, the new instrument allowed the observer to look at and recognise the physical structure of the celestial bodies. In a broader sense, the new instrument also permitted to distinguish the physical structure of the Universe as a whole. Galileo spent the most part of his life trying to convince astronomers and philosophers of the reliability of telescopic observations, and the validity of cosmologic evidence revealed by the new instrument.

Already four hundred years have passed since Galileo's inauguration of this new way of observational astronomy. Throughout this period of time, Galileo celestial discoveries have been celebrated several times with publications, symposia, lectures, lessons, memorial statues, movies, and so forth. The purpose of this exhibition is to demonstrate, by a multi-approach research, the intrinsic difficulties and the complexity of Galileo's operation, and its impact on the society of the time. The examination of the historical context by an international team of historians is merged with the results of the scientific analysis of 17th century lenses conducted by the National Institute of Applied Optics (INOA) in Arcetri, the National Institute of Nuclear Physics (INFN), Departments of Florence and Milan, and the Electronics and Information Department of the Polytechnic School of Milan. Moreover, the production of a faithful replica of the telescope – the tube by J. and R. Morris, the optical components with the scientific consultancy of the Glass Experimental Station in Murano – will allow the members of the Arcetri Astrophysical Observatory to perform new observations of the celestial bodies Galileo had examined. Thanks to this multi-approach research, and the use of multimedia reconstructions, this apparently over investigated chapter of the history of astronomy is revealing new and unexpected insights.

Fig. 0.2 - G. Galilei, Telescope, *c.* 1610, detail (IMSS, inv. 2428).

Fig. 1.1.1 - G. Parigi, *Archimedes' burning mirror*, early 17th C. (Uffizi Gallery, Room of Mathematics, Florence).

Chapter 1:
The Prehistory of the Invention of the Telescope

Sven Dupré

1.0 - Introduction

The dream of telescopic magnification is perhaps even as old as mankind. Already in the 13th century, Roger Bacon (*c.* 1219-*c.* 1292) dreamt of a telescopic device with which "a child might appear a giant, and a man a mountain". However, the device was most useful to see approaching armies, Bacon thought. The projected military usefulness of a telescope was also central to the widely circulated legend of the telescopic mirror which King Ptolemy allegedly had built atop a tower in Pharos to "see by it the enemies ships, that invaded his country, and plundered it", as Giovan Battista Della Porta (*c.* 1535-1615) reminded us in *Magiae Naturalis* (1589).

If telescopic magnification had been an age-old dream of mankind, then what was it that stood between dream and reality? Why did it take until September 1608 for a spectacle-maker in Middelburg to apply for a patent for a telescope? What were the impediments to the invention of the telescope? The answer to this seemingly simple question is complex, and perhaps the question is too complex to be fully and adequately answered by present historical research. Yet, it is possible to point to significant developments in optical theory and the craft of lens-making, especially in the two centuries preceding the invention of the telescope. Such developments are important elements for any possibly satisfying answer to the question of the prehistory of the invention of the telescope.

Optics was a well-established mathematical discipline with a long tradition reaching back to Antiquity. During the Middle Ages the discipline of optics was known as '*perspectiva*', and therefore, historians refer to this tradition as the perspectivist tradition. In the 16th century, mathematicians involved in the design of mirrors and other optical objects grappled to come to terms with the properties of the mirror, the lens, and the *camera obscura*. Some of these mathematicians were well-versed in the perspectivist tradition of optics, and their attempt to understand these optical objects gave way to conceptual innovations (and confusions). One of these concep-

tual innovations was materialized in the description of a telescopic device as early as 1580. This design, however, immediately proved to be a dead end, because it was not adapted to the limitations imposed by the craft of lens-making. Improvements in lens-grinding techniques eventually made possible the invention of a telescope of a different, so-called Galilean design.

1.1 - The Perspectivist Tradition

Optics is a discipline with roots in Antiquity. The works that founded the discipline were those of Euclid (4th century B.C.E.) and Claudius Ptolemy (*c.* 100 – *c.* 178 C.E.). Their optical analyses were based on an extramissionist theory of vision in which visual rays travel from the eye to the objects in the observed world. Together these visual rays form a visual pyramid of which the apex is located in the eye. The primary concern of optics in Antiquity was to explain how we see the world. Euclid's *Optics* was limited to the problem of direct vision. Ptolemy's

Fig. 1.1.2 - Witelo, *Peri Optikes ...* , Nuremberg, 1535: Frontispiece.

Optics also included sections on catoptrics, or the study of reflection upon a mirror-like surface, and dioptrics, or the study of refraction through a surface separating two transparent media. For reflection the law of equal angles (the angle of incidence equals the angle of reflection) became connected with the name of Hero of Alexandria (1st century C.E.). Ptolemy also measured refraction in different media which allowed him to collect tables of refraction in which angles of incidence were connected with angles of refraction.

The arguably most important work in pre-modern optics was *Kitāb al-Manāzir* by the Islamic mathematician Ibn al-Haytham (965-1039), who became known in the West under his Latinized name, Alhazen. The contribution of Ibn al-Haytham was his reconciliation of an intromissionist theory of vision – in which light rays travel from the objects in the observed world to the eye – with the concepts and results of his mathematical predecessors, Euclid and Ptolemy. When Ibn al-Haytham's work was transmitted to the Latin West, it rapidly became the text on which the perspectivist tradition of optics was founded in the 13th century. The perspectivist canon consisted of the works by Bacon, *Perspectiva Communis* by John Peckham (*c.* 1230-1292), and *Perspectiva* by Witelo (*c.* 1225-*c.* 1300). These works reconciled the optical concepts of the mathematicians with the Aristotelian theory of the nature of light.

It is important to realize that, unlike modern geometrical optics, the primary concerns of the perspectivists were to understand vision and visual cognition. As a consequence of this framework, in their discussion about reflection and refraction, the perspectivists were only interested in the perception of images in mirrors and glass spheres. This does not mean that the problems related to the behaviour of light reflected in a mirror and refracted in a glass sphere (or in lenses when eyeglasses became known in the late 13th century) were ignored. In fact, there was a long and thriving tradition of the study of spherical and parabolic burning mirrors that reached back to Antiquity. In the 16th century the study of burning mirrors received considerable impetus from its connection with Archimedes (287-212 B.C.E.) who – in Pliny's version – was said to have burned the enemies' ships at Syracuse **(Fig. 1.1.1)**. This military feat was attractive to mathematicians eager to find patronage for their optical studies. Surprisingly, however, the study of burning mirrors (and the location of the burning point) was considered fully independent from the location of the perceived images in mirrors and glass spheres.

During the 16th century, the texts of the perspectivist canon had become widely available in print. For example, in 1535, Georg Tannstetter and Petrus Apianus printed

Fig. 1.1.3 - F. Risner (ed.), *Opticae Thesaurus...* , Basel, 1572: Frontispiece.

Fig. 1.2.1 – G. Galilei, Copy of *Theorica Speculi Concavi Sphaerici* by E. Ausonio, between 1592 and 1601 (BNCF, Ms. Gal. 83, fol. 4r).

the first edition of Witelo's *Perspectiva* **(Fig. 1.1.2)**. In 1572, Frederic Risner published an edition of Ibn al-Haytham's and Witelo's works together in one book, allowing for cross-referencing between the two works **(Fig. 1.1.3)**. In the same period, interest in the optical works from Antiquity was revived. First Johannes Regiomontanus, and later Georg Hartmann planned an edition of Ptolemy's *Optics*, which failed because of the complexities of having to work with a fragmented manuscript. In 1542, however, Hartmann did succeed to print his edition of Peckham's *Perspectiva Communis* which became the basis of all subsequent editions. Painters' adoption of linear perspective revived interest in Euclid's *Optics*. In 1557, Jean Pena published a highly influential edition of Euclid's *Optics* together with the pseudo-Euclidean *Catoptrics*. Around the middle of the 16th century, optics had not only entered the university curriculum, the canonical texts of the perspectivist tradition of optics had also become widely accessible, including in Italy.

1.2 - Practical Optical Knowledge of Mirrors and Lenses in the 16th Century

Mid-16th century Italy saw the introduction of conceptual innovations in optics by mathematicians and natural magicians, who were well-versed in the perspectivist tradition of optics, but also interested in the design of concave mirrors. It was left to

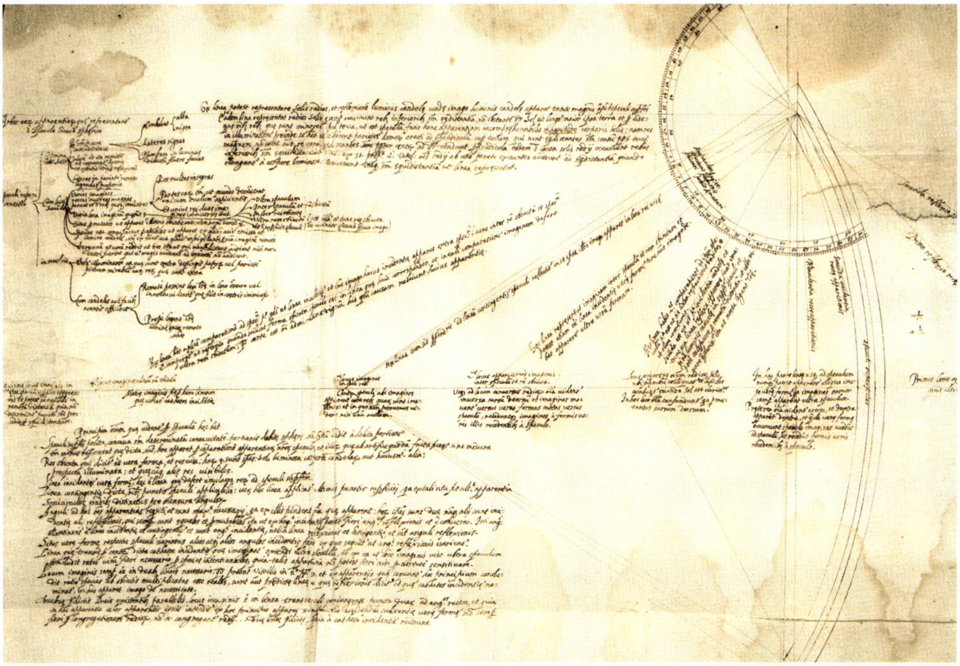

Ettore Ausonio (*c.* 1520-*c.* 1570), a Venetian physician and mathematician famous for his mirrors, to attempt to integrate the problems of the locations of the burning point and the perceived images in a concave mirror, two fields of study which, as we have seen, had always been separated in the optical tradition. In the 1560's – shortly after Della Porta published his first edition of *Magiae Naturalis* (1558) – Ausonio wrote and drew a *Theorica Speculi Concavi Sphaerici*; published in a modified version by the Bolognese astronomer Giovanni Antonio Magini in 1602. However, Ausonio's manuscript version (without Magini's changes) was circulated in the later 16th and early 17th centuries. We know of it today from a handwritten copy by Galileo **(Fig. 1.2.1)**.

In his *Theorica* Ausonio introduced a concept, which later in Della Porta's *De Refractione* (1591) became known as the '*punctum inversionis*'. Ausonio identified the location of the burning point of a concave mirror with its 'point of inversion', the point where the orientation of the image perceived in the mirror changes. Moreover, Ausonio argued, this is also the point where the eye, if located there, perceives a maximally magnified and maximally confused image. It is important to keep this in mind when we will turn to attempts at telescopic magnification. For now, we should realise that the introduction of '*punctum inversionis*' is the first attempt to integrate the study

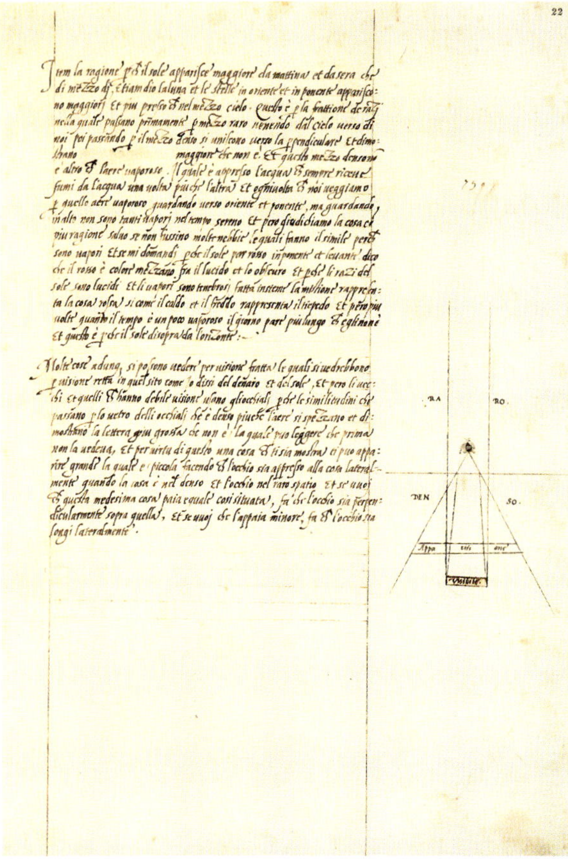

Fig. 1.2.2 - G. Fontana, *Della Prospettiva*, ms., 1400-1450: Eyeglasses (Biblioteca Riccardiana, Florence: Ricc. 2110, fol. 22r).

Fig. 1.2.3 - G. B. della Porta, *De Refractione*, Naples, 1593: Image formation in a convex lens.

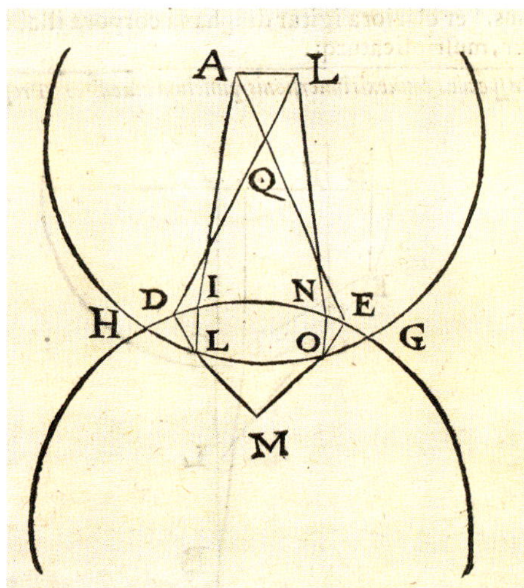

of burning mirrors and of image perception in mirrors within the framework of the perspectivist tradition.

After the invention of eyeglasses with convex lenses to correct presbyopia in the late 13th century, the study of lenses was not ignored. However, the images that were perceived through the lenses were discussed in a way that was disconnected from the burning properties of convex lenses. In a mid-15th century manuscript, attributed to the physician and mathematician Giovanni Fontana (*c.* 1395-*c.* 1455) – that was still used in the classroom of Ostilio Ricci, who also taught Galileo mathematics – eyeglasses were explained with the same principles by which Ptolemy explained the magnification of an object under water **(Fig. 1.2.2)**. It was left to Della Porta to transfer the concept of '*punctum inversionis*' from the study of concave mirrors to convex lenses. In *De Refractione* (1591) he identified the burning point with the point of inversion of a convex lens. He also considered this the locus where the image seen through the lens was maximally magnified **(Fig. 1.2.3)**.

The concept of '*punctum inversionis*' is a conceptual innovation transgressing the boundaries of optical knowledge in the perspectivist tradition. It should be considered practical optical knowledge, not only because it was invented by mathematicians and natural magicians interested in the design of mirrors, but also because it was based on empirical familiarity with the burning and imaging properties of concave mirrors. Moreover, it was even possible to use the concept of '*punctum inversionis*' combining the burning and imaging properties of concave mirrors and convex lenses without being as well-versed in the perspectivist tradition of optics as Ausonio and Della Porta were. It is important to distinguish this body of practical knowledge from the material knowledge of lens-making. While practical optical knowledge transformed the history of optical theory, material knowledge of lens-making led to the invention of the so-called Galilean telescope.

1.3 - Between the 'Camera Obscura' and the Telescope

That the burning point and the point of inversion of a concave mirror and a convex lens were the same was empirically familiar, but conceptually alien to the perspectivist tradition of optics until Ausonio and Della Porta invented the '*punctum inversionis*'. In a similar vein, the projected image inside a *camera obscura* was empirically familiar, but conceptually alien to perspectivist optics. The tradition of perspectivist optics was only interested in the perception of images. Therefore, it was only natural that projected images had no conceptual place in perspectivist optics. In the

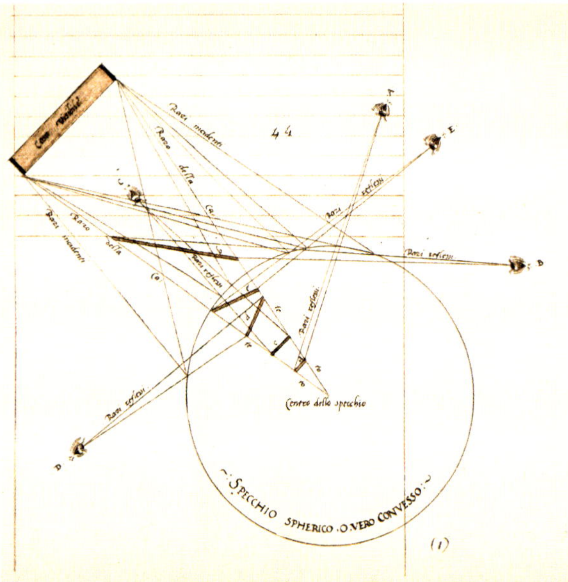

Fig. 1.3.1 - G. Fontana, *Della Prospettiva*, ms., 1400-1450: Image outside a convex mirror (Biblioteca Riccardiana, Florence: Ricc. 2110, fol. 12r).

Fig. 1.3.2 - G. Fontana, *Bellicorum Instrumentorum Liber*, ms., 1420-1440: Magic Lantern (Bayerische Staatsbibliothek, Munich: Cod. Icon. 242, fol. 70r).

15th and 16th centuries, mathematicians and natural magicians sometimes assimilated empirically generated projected images with the images in the geometrical (not the physical) locus outside a convex or concave mirror (which followed from a correct application of a geometrical construction to locate images used in the perspectivist tradition of optics) **(Fig. 1.3.1)**. This assimilation made these mathematicians and natural magicians create (on paper) allegedly experimentally produced images strangely floating in the air. A good example of this conceptual confusion is found in a manuscript, *Bellicorum Instrumentorum Liber* by Fontana (likely author of *Della Prospettiva*). Fontana illustrated a sort of magical lantern which allegedly projected the image of a demon floating in the air to terrify the enemy **(Fig. 1.3.2)**.

It was left to Johann Kepler (1571-1630) to sort out this conceptual confusion in an experiment which he witnessed in the Dresden *Kunstkammer*. His success with the conceptual integration of projected images was one part of the solution to the puzzle which allowed him to launch his new theory of vision in *Paralipomena* (1604), which is still at the basis of our own understanding of the problem today. It is not a coincidence that Kepler's experiment was situated in a *Kunstkammer*. In the 16th century, the images produced by optical devices were often shown to visitors of the princely and other collections. Although Kepler's conceptual integration of projected

Apparentia nocturna ad terrorem
videntium

[illegible]

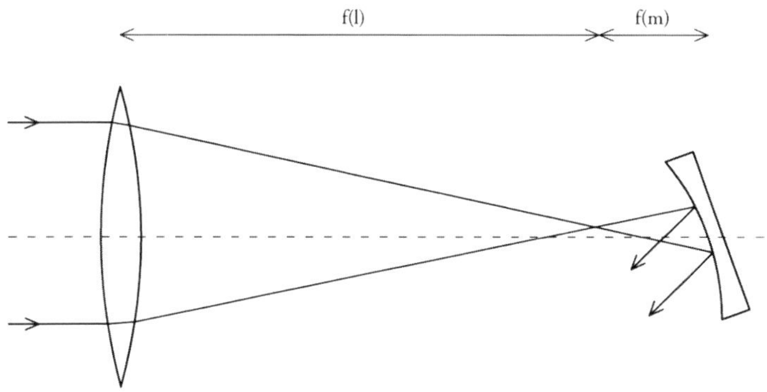

Fig. 1.3.3 - Scheme of William Bourne's telescope design: f(l) = focal length of the convex lens; f(m) = focal length of the concave mirror.

images inside the *camera obscura* was at the basis of the emergence of a new, early modern era in the history of optics, adaptations to the instrument itself – to impress patrons and visitors, and perhaps to aid painters with topographical images – gave way to attempts at telescopic magnification. In *La Pratica della Perspettiva* (1563) Daniele Barbaro (1514-1570) described the effects of a convex lens placed in the aperture of the *camera obscura*. In his Italian translation of Euclid, *La Prospettiva* (1573), the mathematician Egnazio Danti (1536-1586) showed how a plane mirror should be used inside the *camera obscura* to re-invert the image. However, most proficient in this respect was, no doubt, Della Porta, who described several combinations of lenses and mirrors in *Magiae Naturalis* (1589). One of Della Porta's *camera obscuras* consisted of a combination of a convex lens (in the aperture) and a concave mirror. What Della Porta did not realise was that this *camera obscura* was, in fact, a telescope.

Remarkably, in a letter to Lord Burghley, the English mathematician William Bourne (d. 1583) proposed a telescope of the same combination of a convex lens and a concave mirror **(Fig. 1.3.3)**. Bourne argued that the convex lens and concave mirror should be placed at such a distance from each other so "that every glasse doth make his largest beame". Interestingly, Bourne described the properties of the convex lens and the concave mirror in terms of 'burning beams' and 'perspective beams'.

The 'burning beam' indicated the locus of the burning point, while the 'perspective beam' indicated the locus where the image was largest when it filled the complete surface of the lens, which happened near the locus of the burning beam, as Bourne told his potential patron. Bourne, thus, described the properties of the convex lens and the concave mirror of his projected telescopic device in terms which combined the burning and imaging properties of mirrors and lenses. In other words, he described them in terms which we have come to recognize as those of Ausonio's and Della Porta's '*punctum inversionis*', although Bourne was less (if at all) versed in the perspectivist tradition of optics. Bourne's conception of magnification was most significant in illustrating his reliance on practical optical knowledge. It would be natural to think of magnification as dependent upon the diameter of the lens, and Bourne's advice for the diameter of the lens was indeed "the broader the better". Regrettably for Bourne, his reliance on this concept of magnification was also the reason why his device immediately proved to be a dead end. Lens-makers of the period would have found it exceedingly difficult to grind lenses of ever larger diameters. Bourne's design was never heard of again. Such a failure reveals that craft knowledge was essential to the invention of the telescope.

1.4 - The Craft of Lens-making

Eyeglasses with convex lenses to correct presbyopia were invented in Italy around 1285. By the mid-15th century also eyeglasses with concave lenses to correct myopia had become widely available. Eyeglasses to correct presbyopia were available in several age categories from thirty to seventy years; the eyeglasses to correct myopia, which were most likely less in demand, only in two categories.

In the 15th century, Florence was a premier centre for the spectacle-making trade **(Fig. 1.4.1)**. Traces of 52 Florentine spectacle makers – including four friars – between the early 15th and the mid-16th centuries have been found in the archives. Their shops were located in the quarters of St. Giovanni and St. Croce. In the 16th century other centres of spectacle-making emerged, such as Venice, home of a thriving glass industry. However, even in the 15th century, Florence had to compete with German spectacle-making centres in Nuremberg and Regensburg. Eyeglasses produced in these centres were products of international trade with other European countries, including England. In the 15th century, eyeglasses were not luxury goods for the wealthy, but on the contrary, within reach of most people as well as available in most European countries.

In the late 13th century the Venetians invented a new type of glass which was much clearer with only a faint colour and fewer bubbles than the greenish 'wald-glass' typically produced. This type of glass became known as '*cristallo*' – the name referred to rock crystal. Venetians used the '*cristallo*' glass to make luxury glass products, including mirrors, for which Venice became famous. However, in the 16th century, burning mirrors were often still made of metal. The process of their making was discussed in *De la Pirotechnia* (1540) by Vannoccio Biringuccio (1480-1537), and *De Re Metallica* (1556) by Georg Agricola (1494-1555). In the same century the migration of Italian glass-makers – who established glass factories in several European countries – caused a wider distribution of the luxury glass products *à la façon de Venise*. After a visit to the Italian glass factories in Antwerp, Antonio Neri (1576-1614) published the first manual on glass-making, *L'Arte Vetraria* (1612), in Florence **(Fig. 1.4.2)**. However, manuals of this style, including Neri's, were completely silent on the making of mirror glass or spectacle-lenses.

Only recent studies of preserved eyeglasses from the period between the 14th and 16th centuries uncovered improvements in lens-making that explain why the telescope only became possible in the later part of this period from the point of view of

Fig. 1.4.1 - Florentine rivet bone or ivory spectacle frame, late 15th C. (Soprintendenza Archeologica per la Toscana, Florence, inv. 225174)

L'ARTE

VETRARIA

DISTINTA IN LIBRI SETTE

DEL

R. P. ANTONIO NERI

FIORENTINO.

Ne quali ſi ſcoprono, effetti marauiglioſi, & inſegnano ſegreti belliſsimi,

DEL VETRO NEL FVOCO & altre coſe curioſe.

ALL'ILLVST.mo ET ECCELL.mo SIG.

IL SIG. DON ANTONIO MEDICI.

IN FIRENZE.

Nella Stamperia de' Giunti. M. DCXII.

Con licenza de' Superiori.

Fig. 1.4.2 - A. Neri, *L'Arte Vetraria*, Florence, 1612: Title page.

lens-making technology. Early eyeglasses were made by cutting discs from blown glass spheres. It was necessary to grind one side of such discs: the inner concave surface to make a lens for a presbyopic eye, the outer convex surface to make a lens for a myopic eye. Presumably in the late 15th century spectacle-makers in Nuremberg shifted to a different technique. They cut discs from plane window glass. While one surface of the lens remained plane, the other surface was grinded in a concave or convex mould. The technique allowed the spectacle-maker to improve the sphericity of the convex or concave surface. This improvement of the quality of lenses made the invention of the telescope possible for those spectacle-makers who thought of combining better-quality convex and concave lenses.

TROPICO
CCCL
CCCXXXX
CCCXXXV
CCCXXX
CCCXXV

CHAPTER 2:
GALILEO'S TELESCOPE

Paolo Del Santo, Jim & Rhoda Morris, Giorgio Strano, Albert Van Helden

2.0 - Introduction

In late 16th and early 17th centuries, the night sky was perceived as a system of crystalline spheres, eight of which carrying the celestial bodies. The so-called 'fixed' stars – those maintaining their reciprocal positions through time – were placed on the most external of such spheres, and each at the same distance from the immobile Earth. The sphere of the stars rotated once a day from east to west, being responsible for the rising and setting of all the celestial bodies. The known planets were placed on the remaining seven spheres in the order established by Ptolemy in *The Almagest*, from the farthest from to the nearest to the Earth: Saturn, Jupiter, Mars, the Sun, Venus, Mercury and the Moon **(Fig. 2.0.1)**. Each planet rose and settled daily by the rotation of the eighth sphere. In addition, each planet had a proper motion – mainly from west to east – which made it change the position in front of the fixed stars. The role of the astronomer was to measure the celestial co-ordinates of the stars and the planets with very precise instruments from night to night.

From Hipparchus of Nicea (2nd century B.C.E.) to Tycho, the observational instruments of astronomy had undergone minor structural changes. Notwithstanding the particular innovations by the single observer, they usually were made of fixed or rotatable frames carrying graduated scales and a suitable set of sights to aid the naked eye. By using such instruments, the observer might distinguish minimal variations in the position of the celestial bodies. Quadrants, sextants, rings and rulers of considerable size were the instruments that equipped the observatory in Alexandria of Egypt (2nd century C.E.), as well as the Islamic observatories of Marāgha (late 13th century), Samarqand (early 15th century), and Istanbul (mid-16th century), and Tycho's observatories of Uraniborg and Stjerneborg (late 16th century). As a result of this tradition, observational astronomers were constantly updating the numerical data from the past to refine usual planetary models or elaborate better planetary models. This never-ending activity was intended to improve the quality of human condition by

Fig. 2.0.1 - A. Santucci's model of the geocentric cosmos, 1588-1593, (IMSS, inv. 714), detail.

obtaining reliable astrological predictions and by solving time and calendar problems to regulate the rhythm of civil and religious life.

As a consequence of the cultural and astronomical context, it is a matter of fact that, independently from the afore mentioned geocentric structure of the Universe or from the heliocentric and the geo-heliocentric cosmologic alternatives by Copernicus and Tycho respectively, the heavens were considered thoroughly explored. In the early 17th century, all the existing planets were know and studied, and all the existing fixed stars were known and catalogued. Nothing new – with exceptions for the appearance of comets and *novae* – was expected from any careful and systematic observation of the sky. To bring to mind this situation is extremely important in understanding the extraordinary innovative impact that the invention of the telescope and Galileo's introduction of the instrument in astronomical observation had on the culture of the time. (*G.S.*)

Fig. 2.1.1 - Portrait of Hans Lipperhey (P. Borel, *De Vero Telescopii Inventore*, 1656).

Fig. 2.1.2 - Portrait of Sacharias Janssen (P. Borel, *De Vero Telescopii Inventore*, 1656).

2.1 - News about Early Spyglasses

On 25 September 1608, the government of the province of Zeeland in the Netherlands, wrote a letter to one of the province's representatives to the national government, asking him to smooth the way for a spectacle maker from the city of Middelburg, by the name of Hans Lipperhey **(Fig. 2.1.1)**. They wrote that Lipperhey "claims to have a certain device by means of which all things at a very great distance can be seen as if they were nearby, by looking through glasses which he claims is a new invention".

Lipperhey travelled to The Hague and showed his spyglass to Prince Maurice of Nassau, who was the commander in chief of the Dutch armed forces. Maurice showed it to Ambrogio Spinola, the commander of the Spanish forces in the southern Netherlands: who was in The Hague for discussions about an armistice. Before the States General had a chance to discuss Lipperhey's patent application, Maurice had already given the secret away. But how much of a secret was it to begin with? Spectacles – both convex for presbyopia and concave for myopia – had been around in Europe for centuries. Mirrors of various kinds had been known since antiquity. Had no one ever thought of putting a combination of such optical devices together to make a powerful magnifier?

The literature, especially that of the 16th century, is replete with references to lenses and mirrors, and even combinations of lenses. Therefore, it cannot be argued that putting optical components together was the product of genius. Furthermore, there were several instances in which writers specifically mentioned a combination of a convex and concave lens. Giovan Battista Della Porta wrote in his *Magiae Naturalis* (1589): "Concave lenses will make one see most clearly things that are far off; but convex lenses things nearby. With a concave lens you shall see small things far away very clearly, with a convex lens things that are nearer larger but more obscurely. If you know how to fit them both together, you shall see both faraway and nearby things larger and clearly. I have much helped some of my friends, who saw faraway things weakly, and what was nearby confusedly, so that they might see all things clearly". Could there be a clearer statement about the effects of the two types of lenses and the effects of a combination of them? Moreover, Raffaello Gualterotti (1548-1639) wrote in a letter to Galileo dated 24 April 1610: "It is now twelve years since I made an instrument, but not for the purpose of seeing great distances and measuring the stars, but rather for the benefit of a cavalry soldier in joust and warfare [...] But as it seemed to me a feeble thing, I neglected it".

These are tantalizing references. However, if these referred to a telescope, then why did not the instrument spread across Europe starting in 1589, or 1598? For when Lipperhey came forward with his spyglass, in October 1608, news of the instrument spread like wildfire across Europe and beyond, and the instrument itself was not far behind. The States General in The Hague discussed Lipperhey's patent application on 2 October 1608, and as governments are wont to do, they appointed a committee to look into the matter. This committee entered in discussion with Lipperhey: they asked him to make an improved spyglass for them. He was to use rock crystal for the lenses instead of glass (an indication of the poor quality of the glass), and they wanted the instrument to be binocular: for closing one eye while looking through the spyglass they apparently found awkward and inconvenient. For this, Lipperhey was advanced the sum of 300 guilders and promised 600 guilders when the work was finished to their satisfaction. At that time, the States General would decide on his patent application.

But shortly after these discussions had started, there were new developments. The provincial government of Zeeland alerted the States General that another person had come forward who claimed to know the art of making spyglasses. This person, most likely Sacharias Janssen **(Fig. 2.1.2)**, showed the gentlemen a spyglass. In their letter to The Hague, they wrote that they feared there were more of these instruments around – and that in any case the art of making them could not be kept a secret because as soon as it became known that there was such a device, others would try to make it. This would not be difficult, once one had seen the shape of the tube and could infer that it contained lenses.

Around the time this letter arrived in The Hague, there was another petition, this time from Jacob Metius (d. 1628) from the city of Alkmaar. Metius claimed that he had been spending all his spare time for the previous two years to investigate the wonderful effects that could be obtained by means of glasses. He had a similar device that he had initially used for other purposes but had discovered that it 'stretched' vision so that one could see distant things. He had worked hard to improve the device; a demonstration model made with poor glass had been favourably compared with Lipperhey's spyglass by Prince Maurice himself. Metius was confident that he could improve the device and make it more useful for the government, and he asked for a patent for twenty years that would apply to all those who did not have the invention already, and that those violating it would forfeit the instruments and be fined the considerable sum of 100 guilders for each in-

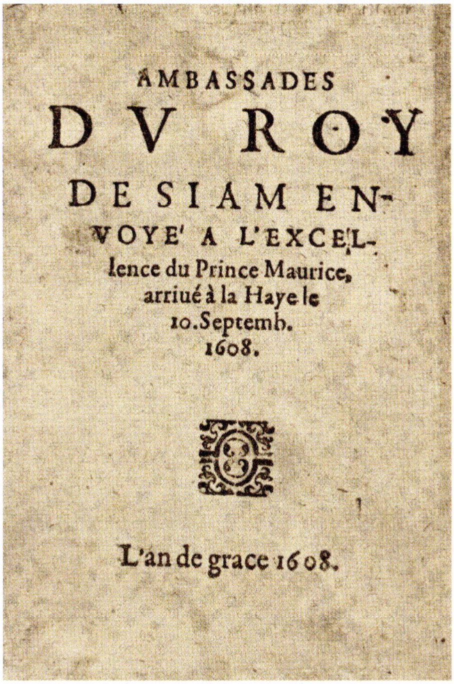

AMBASSADES
DV ROY
DE SIAM EN-
VOYE' A L'EXCEL-
lence du Prince Maurice,
arriué à la Haye le
10. Septemb.
1608.

L'an de grace 1608.

Fig. 2.1.3 - Ambassades du Roy de Siam envoyé à l'Excellence du Prince Maurice, The Hague, 1608: Title page.

stance. This petition was discussed by the States General on 17 October 1608.

What were the gentlemen of the States General now to do? Within three weeks, three different claimants to the invention of this instrument had come forward (although Janssen had not applied for a patent). On the same day that Metius' letter was discussed, the States General awarded him 100 guilders and encouraged him to keep improving his instrument. Lipperhey was busy making a binocular instrument with rock-crystal lenses for the government, and the decision on his patent was still pending. Lipperhey finally finished a binocular instrument early in December 1608. At this point, a decision on his patent application was finally made: it was rejected. Yet, the instrument Lipperhey had made was judged to be of good quality. Of the 600 guilders to be paid, Lipperhey was now given 300, with the assignment of making two more instruments, upon delivery of which, he would be paid the remaining 300 guilders. The record shows that Lipperhey finished this assignment in February 1609 and was paid the remaining 300 guilders.

However, by this time the action had moved far beyond The Hague and Middelburg. Within days of Lipperhey's arrival in The Hague, diplomats were reporting back to their governments about the new device, and they were looking for people who knew how to make it. Some time in October 1608, a newsletter was printed with the title *Ambassades du Roy de Siam envoyé à l'Excellence du Prince Maurice, arrivé à la Haye le 10 Septemb(re) 1608* **(Fig. 2.1.3)**. This was a report of a diplomatic mission from Siam to the Netherlands, but part of the newsletter was devoted to the presentation of Lipperhey's spyglass to Prince Maurice, and it gave a vivid description of the device's power: "From the tower of The Hague, one clearly sees, with the aid of the said glasses, the clock of Delft and the windows of the church in Leiden, despite the fact that these cities are distant from The Hague one-and-a-half, and three-and-a-half hours by road, respectively". And if there was any doubt about its military usefulness, Spinola was reported to have exclaimed: "From now on I could no longer be safe, for you will see me from afar".

As the news spread through diplomatic channels, men with actual spyglasses began to spread out over Europe to try to cash in on them. By the spring of 1609, spyglasses could be bought in spectacle-makers' shops in Paris, and the first examples were already in Italy. In the meantime, a copy of the newsletter had reached

Fig. 2.1.4 - G. B. Della Porta's scheme of the telescope (Letter to F. Cesi, 28 August 1609).

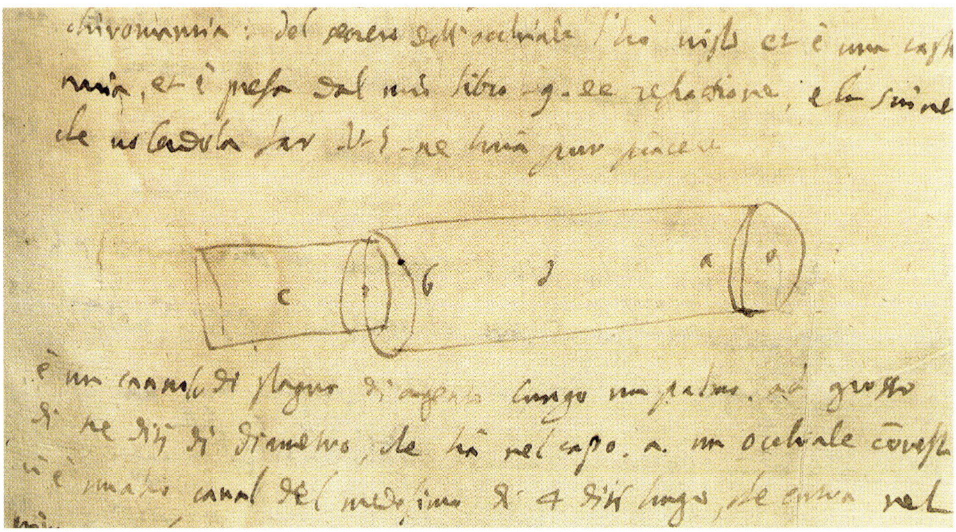

Paolo Sarpi (1552-1623) in Venice in November 1608, sent by one of his correspondents in Paris (and Sarpi was in close contact with Galileo, professor of the mathematical sciences at the nearby University of Padua). The newsletter gave no details about the construction of the spyglass, and so Sarpi wrote to friends in Paris for details. This took several months, but by the spring of 1609, Sarpi and Galileo knew how the spyglass was constructed.

Meanwhile, when in August 1609 a spyglass was brought to Naples, Della Porta looked at it and wrote: "About the secret of the spectacles, I have seen it, and it is a hoax, and it is taken from the ninth book of my *De Refractione*". He went on to describe the instrument pictured here **(Fig. 2.1.4)**. Why after hearing the hoopla preceding the instrument would he call it 'hoax' when he finally saw one? And why would Gualterotti write to Galileo six months later that he had had something like it in 1598 but had considered it a 'feeble thing'? Were they perhaps both right? Could it be that combinations of concave and convex lenses were present in a number of places in Europe among scholars studying the effects of lenses and craftsmen who made them? Evidence of this is mounting. For instance, Metius had such a combination, and the son of Janssen stated twenty-five years later that his father had made the first spyglass in the Netherlands after one in the hands of an Italian and which bore the date 1590.

It appears that the telescope, as we call it, was there before anyone knew it, that combinations of concave and convex lenses were obvious and common. Why, then, did others such as Della Porta not anticipate Lipperhey? It must have been a matter not of the idea or the combination, but of the material. How well were lenses ground and polished? Historians cannot answer that question. But there are a number of early lenses in museums, and opticians can test these and perhaps find out what improvements in glass quality, or grinding and polishing techniques, or some other aspect made the telescope possible. One thing is certain: once it was there, the telescope improved by leaps and bounds, and in less than two years, it revolutionized astronomy.

However, 17th century Europe would not have been satisfied with the conclusion that the telescope was there before anyone knew it. Local pride and scientific curiosity about the 'first inventor' of the telescope led the city father of Middelburg to hold hearings on the matter half a century after the fact. Testimony was heard from surviving relatives and acquaintances of Lipperhey and Janssen. The witness' statements were recorded and eventually published in *De Vero Telescopii Inventore* (1656), a book on the subject by the French physician Pierre Borel (*c.* 1620-1671), in which he declared Janssen as the first inventor. However, this did not lay the issue to rest. In 1682, the famous Dutch physicist and astronomer Christiaan Huygens (1629-1695) obtained a copy of Metius' petition, certified by a notary and member of the Metius family. The claim of Metius was thus resurrected. Since then, claims for all three have been defended. Of course no definitive answer to the question can be given: perhaps it is the wrong question. (*A.V.H.*)

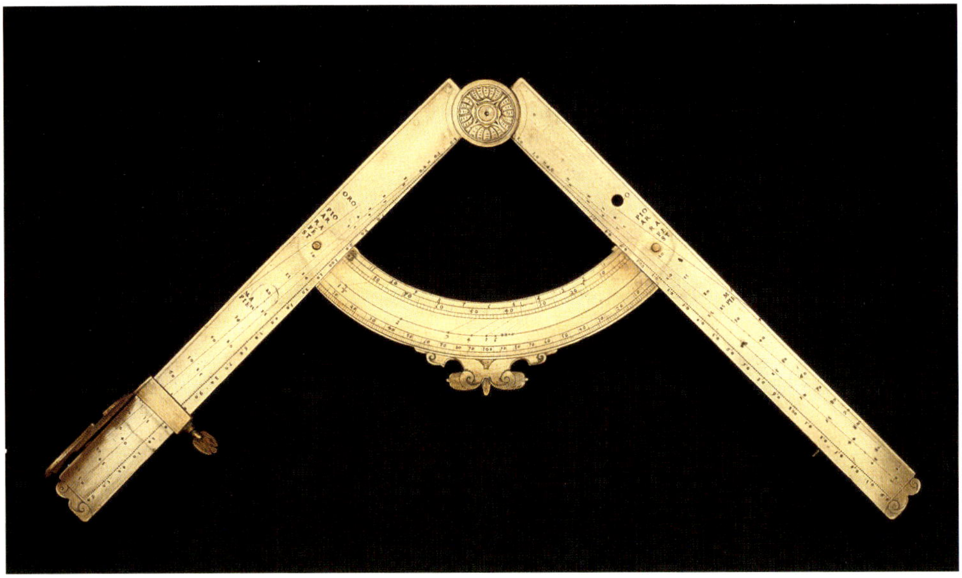

Fig. 2.2.1 - G. Galilei, Geometric and military compass, c. 1606 (IMSS, inv. 2430).

Fig. 2.2.2 - G. Galilei, Armoured magnets, c. 1600 (IMSS, inv. 6-11).

2.2 - Galileo, the Instrument Maker

In 1592, Galileo moved from the University of Pisa to the University of Padua, to the position of "lecturer of mathematics". Placed within the Republic of Venice, Padua was a more active and open-minded city than Pisa. Not to be overlooked, the new university offered Galileo a higher salary, that would have considerably increased every two years.

Since it was not far from Padua, Galileo frequently travelled to Venice in order to observe the naval construction at the Arsenal, and to meet with distinguished friends, such as Giovanfrancesco Sagredo and Paolo Sarpi. Galileo found the situation very favourable. He also established a relation with Marina Gamba, and, between 1600 and 1606, he had with her two daughters and a son. However, the necessity to maintain his family (including his widowed mother and the family of his brother Michelangelo) forced Galileo to try to increase his financial standing. In a few years, he transformed his home into a hostel for up to twenty wealthy students, giving them private lessons on mathematics. Moreover, he established a workshop adjoined to his study, and assumed a specialised craftsman, Marcantonio Mazzoleni, to create mathematical instruments. Both endeavours created a new source of income for Galileo.

In particular, around 1597, Galileo conceived a new type of geometrical and military compass which he produced, or had produced in several copies **(Fig. 2.2.1)**. This compass is the first surviving material testimony of Galileo's interest and ability in designing, improving and handcrafting instruments for both scientific and practical purposes. The compass could be employed to make mathematical and geometrical calculations, as well as used for measuring angles useful in surveying, cartography, ballistics, and even astronomy. As was done, Galileo wrote a manual on *Le operazioni del compasso geometrico et militare* (1606), which also safeguarded him against any form of plagiarism.

The compass was not the only instrument that Galileo conceived and made in Padua. After William Gilbert (1544-1603) published *De magnete* (1600), Galileo was intrigued by the peculiarity of magnetic force. He searched for suitable pieces of magnetite and tried to determine how their intrinsic power could be increased. Within a couple of years, he learnt how to cut, armour and transform pieces of the mineral into powerful magnets **(Fig. 2.2.2)**. Successively, he sent some of them to Sagredo, and planned to sell another to the Grand Duke of Tuscany. Moreover, between 1602 and 1609, Galileo conducted studies on mechanics, and, in particular,

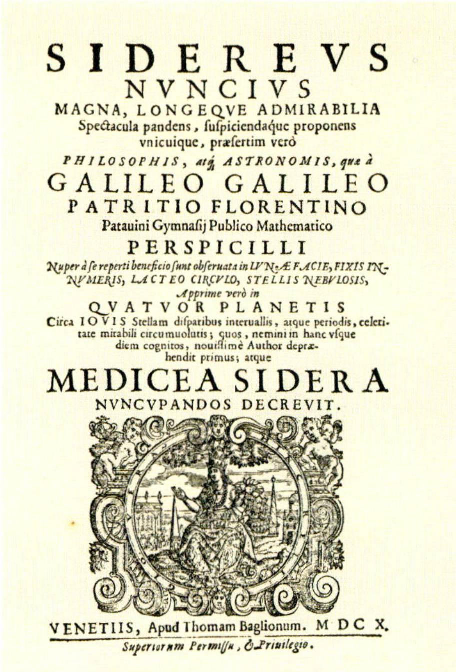

SIDEREVS
NVNCIVS
MAGNA, LONGEQVE ADMIRABILIA
Spectacula pandens, suspiciendaque proponens
vnicuique, præsertim verò
PHILOSOPHIS, atq; ASTRONOMIS, quæ à
GALILEO GALILEO
PATRITIO FLORENTINO
Patauini Gymnasij Publico Mathematico
PERSPICILLI
Nuper à se reperti beneficio sunt obseruata in LVNÆ FACIE, FIXIS IN-
NVMERIS, LACTEO CIRCVLO, STELLIS NEBVLOSIS,
Apprime verò in
QVATVOR PLANETIS
Circa IOVIS Stellam disparibus interuallis, atque periodis, celeri-
tate mirabili circumuolutis; quos, nemini in hanc vsque
diem cognitos, nouissimè Author depræ-
hendit primus; atque
MEDICEA SIDERA
NVNCVPANDOS DECREVIT.

VENETIIS, Apud Thomam Baglionum. M DC X.
Superiorum Permissu, & Priuilegio.

Fig. 2.2.3 - G. Galilei *Sidereus Nuncius*, Venice, 1610: Title page.

on the mathematical law governing the motion of the falling bodies. Once again, in order to accomplish reliable experiments, he made suitable apparatus, including a sort of water clock to measure short time intervals.

By constantly dealing with the material aspect of science – the making and improving of instruments – Galileo took an immediate interest in the news about the optical device from the Netherlands. In December 1608, Sarpi received information on the spyglass, and began to write about it with his correspondents. These included Jacques Badouère in Paris, one of Galileo's students at Padua in the late 1590's. It is uncertain whether Galileo received news about the spyglasses from Badouère or Sarpi, as well as it is uncertain whether or not he had a spyglass from Paris or from Venice. In a letter to his brother-in-law Benedetto Landucci and in *Sidereus Nuncius* (Starry Messenger) (1610), Galileo only refers that, in June 1609, Badouère confirmed the optical properties of the instrument. This was enough for Galileo to begin a careful study of the optical composition permitting the magnification of far objects.

His ability in handcrafting instruments made it possible that, within a few weeks, Galileo grasped the secret of the spyglass. At first, he made one by taking a lead tube, and inserting at the extremities two spectacle lenses, a plano-convex for the objective and a plano-concave for the ocular. This device magnified far objects by three times. However, Galileo was immediately able to improve the instrument; it is unclear whether on the bases of the principles of optics, as he claimed, or on the bases of an empirical process. Essentially, he understood that, in order to obtain higher magnification, it was indispensable to couple a weaker convex lens for the objective with a stronger concave lens for the ocular. The new instrument that Galileo made by following this optical principle had a magnification power of about eight or nine times. Such an instrument was highly superior to any spyglass from the Netherlands, and, on 21 August 1609, Galileo went to Venice to impress a group of noblemen with a demonstration of it from atop the bell tower of St. Mark. Three days after, he donated the instrument to the Venetian Doge, Leonardo Donato, singling out its great military advantage for the navy and the army. By looking through the instrument, enemy forces approaching could have been identified and evaluated hours before their arrival, signifying a valuable advantage for the Venetians. As a result, the government conferred Galileo a life long position at the University of Padua, with an annual salary of 1,000 florins.

To develop the instrument ever further, Galileo needed objective and ocular lenses with an optical quality that no spectacle makers could accomplish. Therefore, he managed to produce lenses in his own workshop, and gave instructions to glass makers from Venice to furnish him with very good pieces of glass. Moreover, to reduce the effect of an improper shaping of the edges of lenses, Galileo discovered the advantage of reducing the aperture of the objective by using a diaphragm. Finally, in early 1610, he had at his disposal sets of very fine lenses to construct an instrument magnifying far objects up to thirty times. This was the instrument Galileo used to perform the revolutionary astronomical observations he described in *Sidereus Nuncius*.

News of Galileo's ability in making the new optical device reached also the Grand Duke of Tuscany, Cosimo II. On 29 August 1609, he informed Galileo about his great interest in the instrument, and the desire to have one. Galileo dedicated the *Sidereus nNuncius* to Cosimo with the aim to come back to Tuscany, and be in the service of one of the most powerful families of the time (**Fig. 2.2.3**). On 19 March, he sent the Grand Duke a copy of the booklet, and the telescope he had used for his

Fig. 2.2.4 - G. Galilei, Objective lens of the telescope sent to Cosimo II in March 1610, (IMSS, inv. 2429).

Fig. 2.2.5 - G. Galilei, Presentation telescope for Cosimo II, c. 1610 (IMSS, inv. 2428), detail.

observations. This was an instrument made just for use and of no aesthetic value **(Fig. 2.2.4)**. However, Galileo could have provided Cosimo with better and more beautifully decorated instruments **(Fig. 2.2.5)**. The Grand Duke appreciated both the dedication and the present. He appointed Galileo the positions of 'first mathematician' at the University of Pisa – with a special exemption from lecturing – and as his 'first mathematician and philosopher', with a salary of 1,000 *scudi* per year.

In a few weeks from the printing of *Sidereus Nuncius*, Galileo's fame spread all over Europe. In particular, the magnification power of his instruments made spyglasses from the Netherlands obsolete. This fact is apparent in the decision by Prince Federico Cesi (1585-1630), founder of the Roman *Accademia dei Lincei*, to give Galileo's instrument a new scientific name, 'telescope', by combining two Greek words: *tele* (at a distance), and *skopeo* (I observe). For several years, the telescope became Galileo's leading product. He furnished wealthy and noblemen, as well as natural philosophers (having a desire to perform astronomical observations) with his telescopes. Paradoxically, he also made telescopes for several individuals who attempted to diminish the importance of the invention, or to credit it to themselves.

The telescope was not the end-point of Galileo's instrument creation. The necessity to carry on reliable astronomical observations, caused Galileo to improve

Fig. 2.2.6 - E. Porcellotti, Reconstruction of Galileo's pendulum time keeper, 1879 (IMSS, inv. 3450).

upon the telescope and transform it into a precision instrument. The so-called 'micrometer' that he used for his observations since January 1612 – perhaps a graduated ruler sliding along the tube of the telescope, as Giovanni Alfonso Borelli (1608-1679) mentioned in *Theoricae Mediceorum Planetarum* (1666) – had precisely this function. In addition, Galileo created a series of paper devices useful in interpreting the numerical data gathered with the aid of the telescope and the micrometer. This was the function of the so-called 'jovilabe', a type of planetary scheme that Galileo prepared between 1610 and 1613. Notwithstanding the originality of their subject – Jupiter's system – the instrument was operated by threads, a method already used in paper instruments for traditional mathematical astronomy.

Finally, besides the idea to use the telescope for military purposes, Galileo also tried to solve with it one of the most important practical problems of his time: longitude. Since 1617, he studied the possibility to determine longitude at sea by using an instrument of his own creation, the '*celatone*'. This was a special helmet with a telescope. The '*celatone*' would aid mariners to determine time differences – and therefore differences in longitude – by careful observation of the position of Jupiter's satellites. After the failure of the negotiation with the Spanish government to have this technique accepted, Galileo re-examined the problem in 1637. He prepared an even more sophisticated technical solution that he submitted to the States General of the Netherlands. In addition to the telescope and the helmet, a special movable platform to insulate the observer from the pitching and rolling of the ship, and a mechanical clock were added. The latter represented the first attempt to improve upon usual mechanical time keepers with a new type of regulating system governed by the isochronal oscillations of a pendulum **(Fig. 2.2.6)**.

In conclusion, the ability and talent as a craftsman was only part of Galileo's multifaceted creativity. Usually, Galileo's inventions and his perfecting of instruments were of great significance for their practical application or their usefulness in the theoretical interpretation of the world. Due to this, the accurate examination of Galileo's surviving telescopes is indispensable for understanding the material basis of his research, but not yet its revolutionary implications for astronomy. (*G.S.*)

2.3 – Galileo's Observations

In March 1610 the *Sidereus Nuncius* was printed in Venice by the publisher Tommaso Baglioni. In it Galileo announced to the world his first, extraordinary astronomical discoveries made with the telescope. This short treatise met with resounding success. The initial print run of 550 copies sold out rapidly, and the echo of Galileo's discoveries resounded all over Europe. In the following months he made other, but no less important discoveries, destined to change the traditional vision of the cosmos forever and to launch that revolutionary process of unification through which the heavenly bodies were subjected to the same laws obeyed by the terrestrial ones. All of this shows not only the unquestionable superiority of Galileo's telescopes to those of his contemporaries, but also and above all his capacity to see the new, where the advocates of traditional culture had recurred to devising complex hypotheses in the attempt to salvage Aristotelian cosmology.

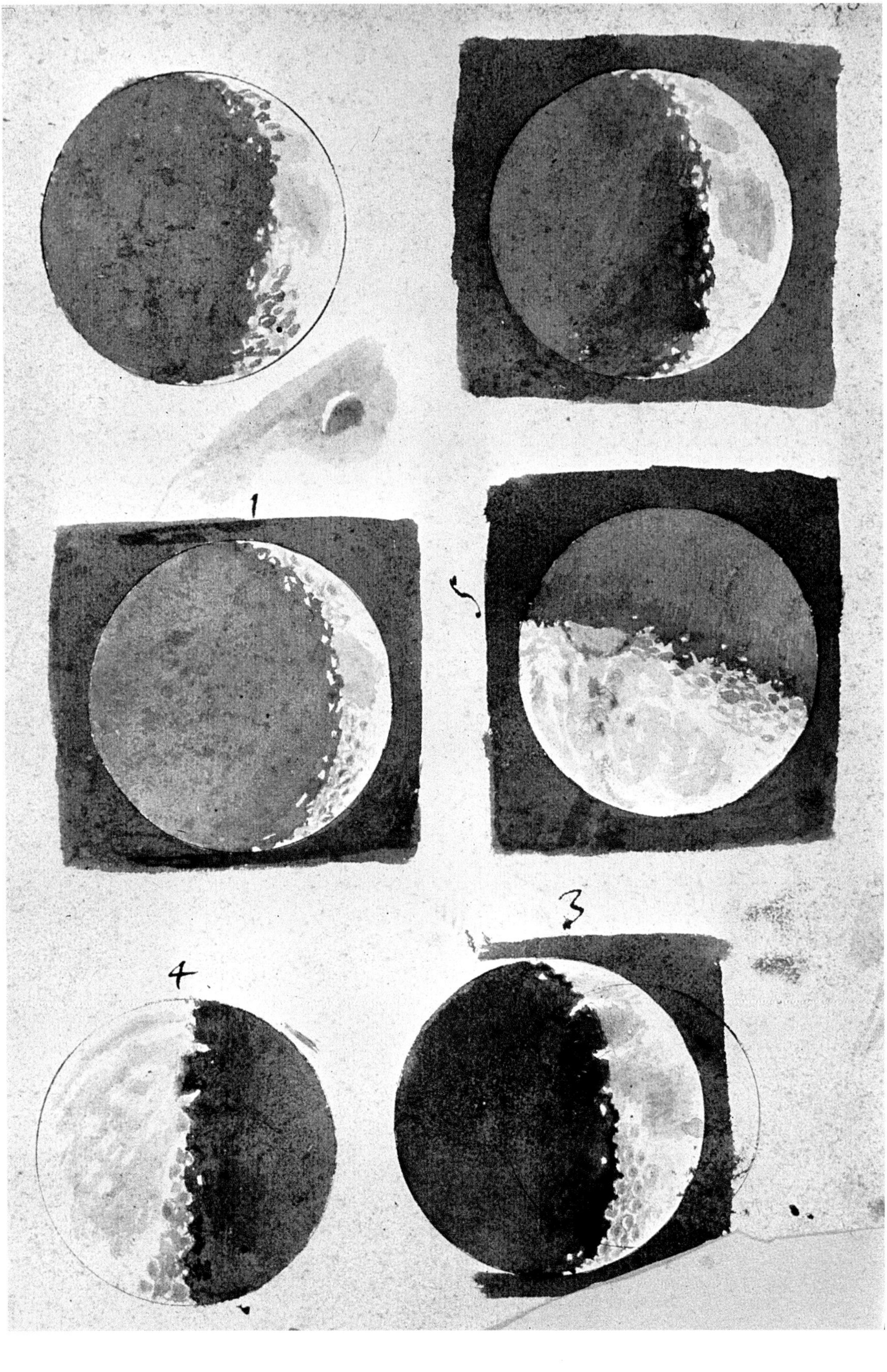
1
3
4

The Moon: Already Thomas Harriot (*c.* 1560-1621), preceding Galileo in the astronomical use of the telescope, had observed the Moon with an instrument of 6-power magnification on July 26, 1609 (according to the Julian calendar, corresponding to August 5 in the Gregorian calendar, at the time not yet adopted in England). But comparison between the drawing of his observation made by Harriot and Galileo's tables of the Moon, which were moreover of striking pictorial effect **(Fig. 2.3.1)**, immediately revealed the superiority of Galileo's optics.

By means of the telescope Galileo arrived at the "conviction that the surface of the Moon is not smooth, uniform, and precisely spherical as a great number of philosophers believe it (and the other heavenly bodies) to be, but is uneven, rough,

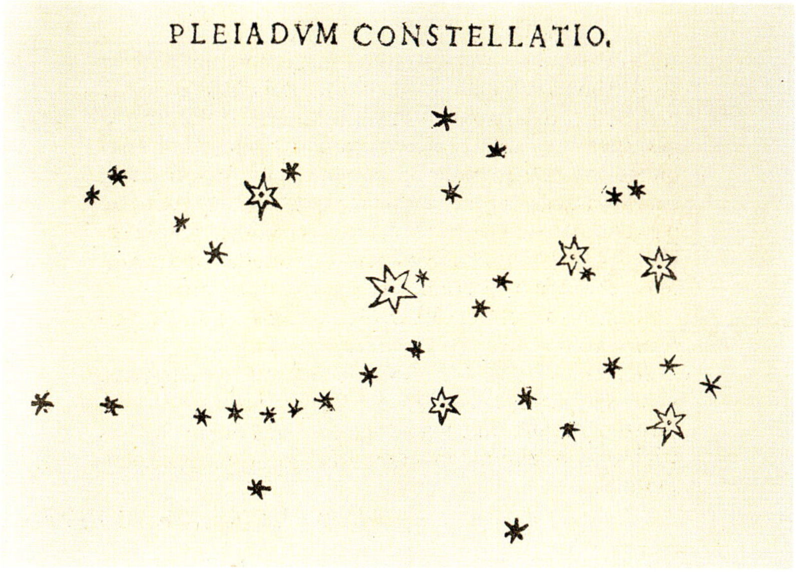

Fig. 2.3.1 - G. Galilei, Water-colour of the Moon seen through the telescope (BNCF, Ms. Gal. 48, c. 28r).

Fig. 2.3.2 - The Pleiades seen through the telescope (G. Galilei, *Sidereus Nuncius*, 1610, c. 16r).

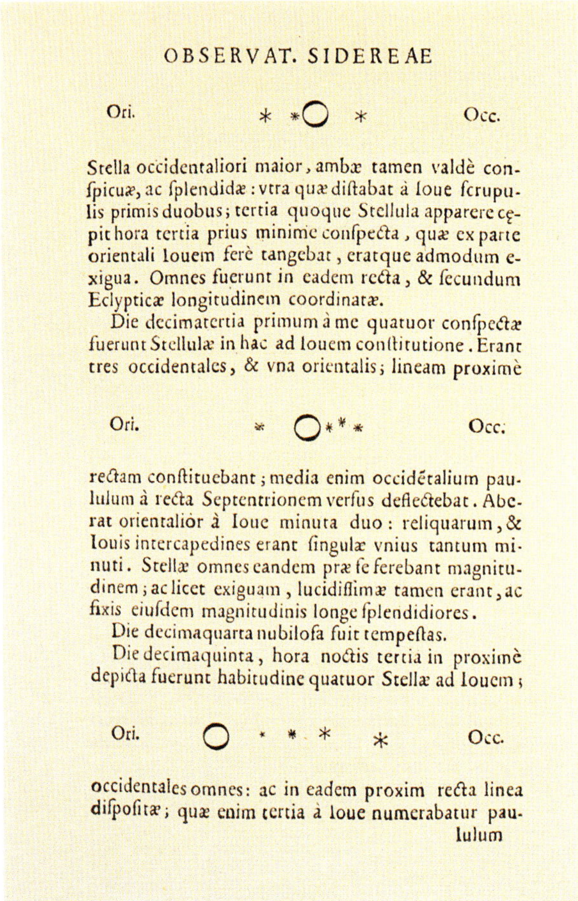

OBSERVAT. SIDEREAE

Ori. Occ.

Stella occidentaliori maior, ambæ tamen valdè conſpicuæ, ac ſplendidæ: vtra quæ diſtabat à Ioue ſcrupulis primis duobus; tertia quoque Stellula apparere cępit hora tertia prius minime conſpecta, quæ ex parte orientali Iouem ferè tangebat, eratque admodum exigua. Omnes fuerunt in eadem recta, & ſecundum Eclypticæ longitudinem coordinatæ.

Die decimatertia primum à me quatuor conſpectæ fuerunt Stellulæ in hac ad Iouem conſtitutione. Erant tres occidentales, & vna orientalis; lineam proximè

Ori. Occ.

rectam conſtituebant; media enim occidẽtalium paululum à recta Septentrionem verſus deflectebat. Aberat orientaliòr à Ioue minuta duo: reliquarum, & Iouis intercapedines erant ſingulæ vnius tantum minuti. Stellæ omnes eandem præ ſe ferebant magnitudinem; ac licet exiguam, lucidiſſimæ tamen erant, ac fixis eiuſdem magnitudinis longe ſplendidiores.

Die decimaquarta nubiloſa fuit tempeſtas.

Die decimaquinta, hora noctis tertia in proximè depicta fuerunt habitudine quatuor Stellæ ad Iouem;

Ori. Occ.

occidentales omnes: ac in eadem proxim recta linea diſpoſitæ; quæ enim tertia à Ioue numerabatur paululum

Fig. 2.3.3 - Observations of the 'Medicean Stars' (G. Galilei, *Sidereus Nuncius*, 1610, c. 18v).

and full of cavities and prominences, being not unlike the face of the Earth, relieved by chains of mountains and deep valleys". This constituted a fact of extraordinary importance, entirely irreconcilable with Aristotelian thought, dominant at the time, which held the heavenly bodies to be non-generated, incorruptible and perfectly spherical. It is no coincidence that the Jesuit fathers of the Collegio Romano, consulted in April of 1611 by Cardinal Roberto Bellarmino, confirmed all of the astronomical discoveries made up to then by Galileo, while expressing strong doubts instead as to the roughness of the lunar surface.

The fixed stars and the Milky Way: Before the end of 1609, Galileo had turned his telescope toward the fixed stars as well, discovering that "in addition to the stars of sixth magnitude" – that is, those most faintly visible to the naked eye, in part already catalogued by Ptolemy – there could be seen "with the telescope such a great number of others, invisible to natural sight, as almost not to be believed". As demonstration of this, he included two drawings in the *Sidereus Nuncius*: one of the Pleiades with 30 additional stars invisible to the naked eye **(Fig. 2.3.2)**; the other of the zone of Orion's belt and sword which, in addition to the already known stars, showed 80 new ones. The stars revealed by the telescope were so numerous that Galileo gave up his original intention of representing the entire constellation of Orion because "there are in it [...], disseminated around those already known, within the limits of one or two degrees, more than five hundred stars".

The telescope also allowed Galileo to reveal the true nature of the Milky Way: "Through the telescope we can see so clearly that all of the discussions, for so many centuries the anxious concern of philosophers, are dissipated by the certainty of perceived experience, and we are freed from sterile disputes. The galaxy is, in fact, no other than a mass of innumerable stars", which cannot be resolved individually by the naked eye.

Amplifying enormously our "natural faculty", the telescope showed, for the first time in history, the existence of a world until then totally unknown and unimaginable, of an entire and much more 'abundant' reality than the one perceptible through the limitations of the senses.

The satellites of Jupiter: On the night of January 7, 1610, while observing Jupiter with "an excellent instrument", Galileo saw in the vicinity of this planet three "small but very bright" stars. The next evening, finding the starlets in a different position in respect to the planet, as if it had moved toward the east – while during those days, according to the predictions, it should have moved to the west – he decided to observe the phenomenon systematically. On the night of the 10th, Galileo suddenly understood that the observations could be explained only by admitting that the three starlets moved around the planet. Then on the night of the 13th, he discovered a fourth little star, it too revolving around Jupiter.

Galileo considered this discovery to be the most important subject presented in the *Sidereus Nuncius*, a discovery that would procure him universal fame. The new stars – which he called '' in honour of the Medici family **(Fig. 2.3.3)** – constituted a proof in favour of the Copernican system, or rather, they eliminated what some had

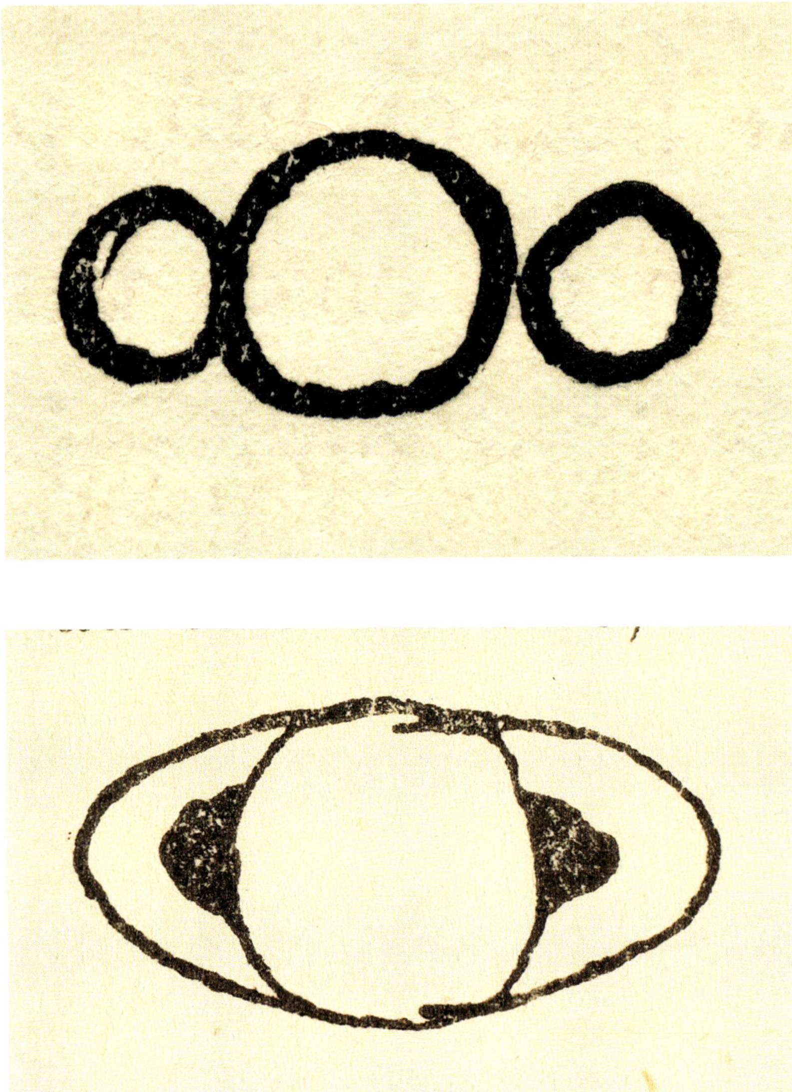

a

b

Fig. 2.3.4 - Saturn formed of three bodies (*a*) and with appendices of semi-elliptical shape (*b*).

thought to be an anomaly in the heliocentric architecture. If in fact, for Copernicus, the Sun was the centre of all of the planetary motions, why was it that only the Earth had a satellite, the Moon, orbiting around it? The satellites of Jupiter demonstrated, on the contrary, that other planets too could have satellites rotating around them.

Three-bodied Saturn: On July 30, 1610, in a letter to Belisario Vinta, Secretary of State under Cosimo II, Galileo reported that he had "discovered another most amazing marvel [...], that the star of Saturn is not one alone, but a combination of three, almost touching one another, which never move nor change among themselves; and they are positioned in line [...], the one in the middle being about three times larger than the other two lateral ones" **(Fig. 2.3.4a)**. Galileo begged Vinta to keep this news secret until it had been published in his planned second edition of the *Sidereus Nuncius*. Subsequently, Galileo also informed Giuliano de' Medici, Tuscan ambassador to Prague, of this discovery, but in the form of a cryptogram.

Two years later, however, a new fact occurred. Saturn's two companions had disappeared, and the planet appeared "solitary". This disconcerting phenomenon cast doubt on all of the preceding observations and was highly disturbing to Galileo, who, in informing Marcus Welser of this news in a letter dated December 1, 1612, arrived at the point of wondering whether it had been "illusive and fraudulent, the appearance with which the crystals [that is, the lenses] have for so long a time deceived" him, as well as others who had observed the two presumed satellites of this planet many times.

Starting from 1616, the appearance of the planet changed again. In place of the two original satellites there were now two lateral appendices in the shape of semi-ellipses, darker at the centre **(Fig. 2.3.4b)**. In short, the true nature of Saturn remained a mystery for Galileo and his contemporaries.

The phases of Venus: In a letter to Giuliano de' Medici dated December 11, 1610, Galileo announced, again by means of an anagram, another sensational discovery. He revealed the solution to the enigma in a letter to Giuliano dated January 1, 1611: "*Cynthiae figuras aemulatur mater amorum*" (The mother of love [Venus] emulates the configurations of Cynthia [the Moon]). In other words, Galileo had discovered that the planet Venus, under telescopic observation, presented phases similar to those of the Moon. **(Fig. 2.3.5)**.

This discovery had far-reaching cosmological implications. In the Ptolemaic system, each planet was thought to move around the immobile Earth along a circumference, called 'epicycle', whose centre rotated in turn along a larger circumference, called 'deferent'. Moreover, the epicycles of Mercury and Venus should always have been found below the Sun, and their centres always located on the line conjoining the Earth and the Sun. Consequently, both of the planets would have shown very reduced phases, never appearing illuminated except in a very thin crescent **(Fig. 2.3.6a)**. In the Copernican system instead – in which the planets, including the Earth, rotate around the Sun, situated at the centre of the Universe – Mercury and Venus would show the entire range of phases **(Fig. 2.3.6b)**. This discovery thus dealt a death blow to the Ptolemaic architecture of the Universe, and as such constituted a crucially significant point in Galileo's most important cosmological work, the *Dialogo sopra i due massimi sistemi del mondo tolemaico, e coperni-*

Fig. 2.3.5 - The phases of Venus (G. Galilei, *Il Saggiatore*, 1623, p. 217).

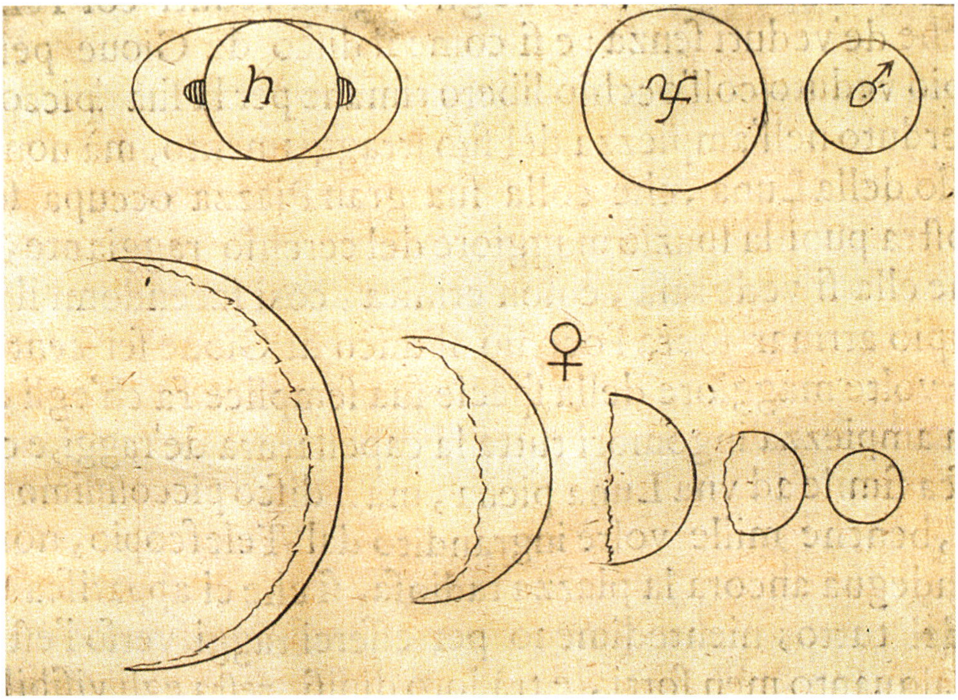

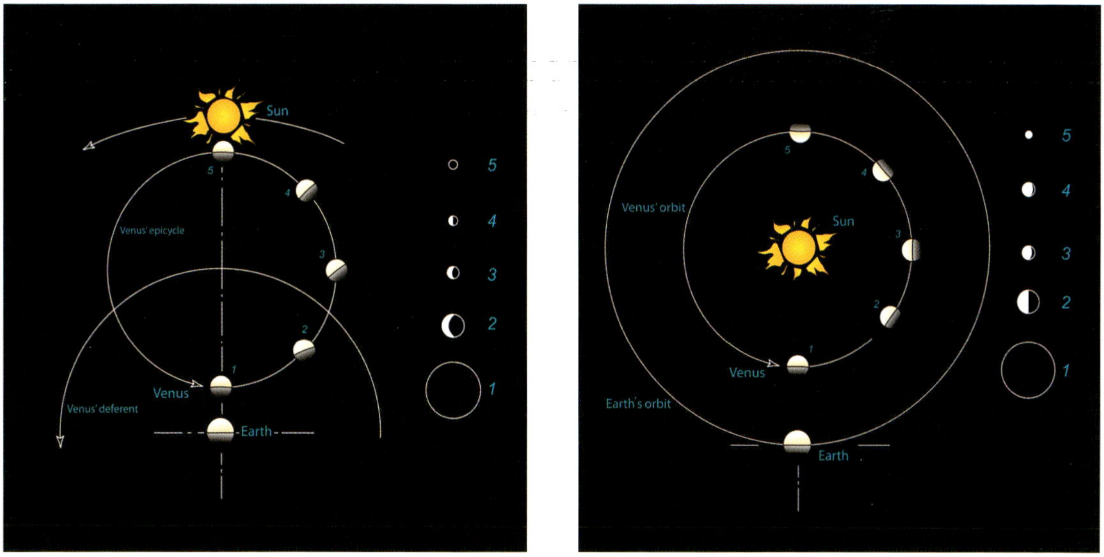

Fig. 2.3.6 - Diagrams of the phases of Venus: *a*) in the Ptolemaic system, *b*) in the Copernican system.

cano (Dialogue concerning the two chief world systems, the Ptolemaic and the Copernican) (1632). In reality the discovery did not constitute an irrefutable proof of the heliocentric thesis, since the system postulated by Tycho around 1584 – according to which all of the planets rotated around the Sun which, as in the Ptolemaic system, rotated around the immobile Earth – provided for phases of Venus (and Mercury) identical to the ones observed by Galileo.

Sunspots: The surface of the Sun, when observed through the telescope, shows dark spots, whose nature (as well as the priority of the discovery) was the subject of bitter controversy between Galileo and the Jesuit father Christoph Scheiner (1573-1650), who in January 1612, under the pseudonym of *Apelles latens post tabulam* (Apelles hidden behind the painting), had published a short treatise called *Tres epistolae de maculis solaribus scriptae ad Marcum Welserum* (Three letters on sunspots written to Marc Welser). To explain the phenomenon of sunspots while at the same time salvaging the Aristotelian dogma of the perfection and immutability of the heav-

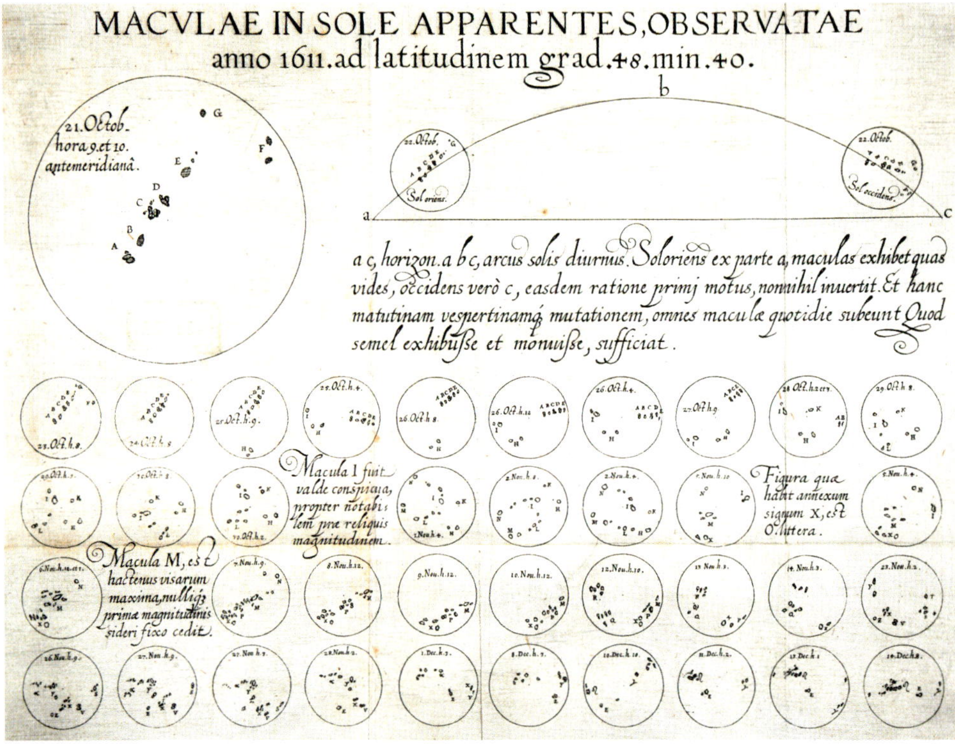

Fig. 2.3.7 – Sunspots (C. Scheiner, *Tres epistolae de maculis solaribus*, 1612.)

ens, Scheiner had hypothesised the existence of clusters of small planets orbiting around the Sun, which, interposed between it and the Earth, appeared as dark spots against the background of the solar disc **(Fig. 2.3.7)**.

In reply to Scheiner's treatise, between May and December of 1612 Galileo sent Welser three letters, which were then published in Rome in the summer of the following year under the aegis of the Academy of the Lincei, with the title of *Istoria e dimostrazioni intorno alle macchie solari e loro accidenti* (History and demonstrations concerning sunspots and their properties) **(Fig. 2.3.8)**. Unlike Scheiner, Galileo believed that the phenomenon took place on the surface of the Sun or in its immediate vicinity, and that the motion of the spots was due to the rotation of

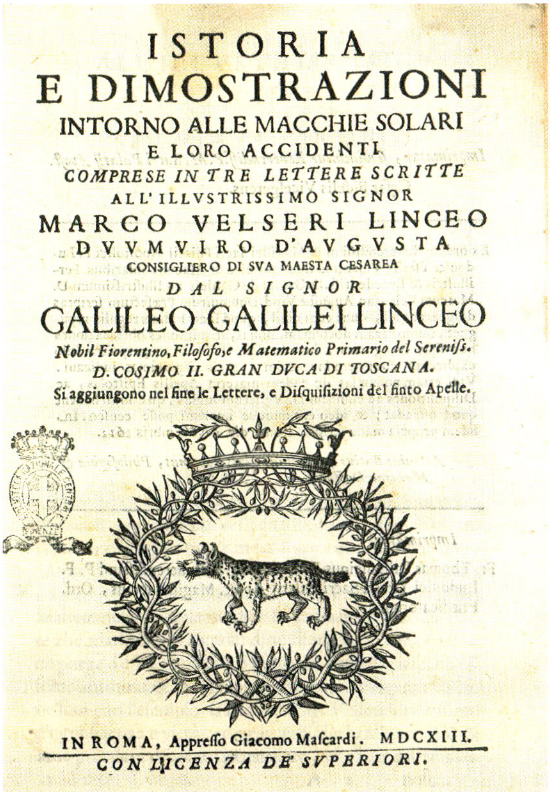

ISTORIA
E DIMOSTRAZIONI
INTORNO ALLE MACCHIE SOLARI
E LORO ACCIDENTI
COMPRESE IN TRE LETTERE SCRITTE
ALL'ILLVSTRISSIMO SIGNOR
MARCO VELSERI LINCEO
DVVMVIRO D'AVGVSTA
CONSIGLIERO DI SVA MAESTA CESAREA
DAL SIGNOR
GALILEO GALILEI LINCEO
Nobil Fiorentino, Filosofo, e Matematico Primario del Serenifs.
D. COSIMO II. GRAN DVCA DI TOSCANA.
Si aggiungono nel fine le Lettere, e Difquifizioni del finto Apelle.

IN ROMA, Appreffo Giacomo Mafcardi. MDCXIII.
CON LICENZA DE' SVPERIORI.

Fig. 2.3.8 - G. Galilei, *Istoria e dimostrazioni intorno alle macchie solari*, Rome, 1613: Title page.

the Sun on its own axis, with a period of about one month. To back up this hypothesis Galileo cited numerous proofs resulting from observations. Sunspots showed no periodicity, appearing and dissolving continuously, assuming mostly irregular shapes and changing from day to day. And as for their presumed opacity and darkness, cited by Scheiner in support of his thesis, Galileo had brilliantly guessed that the sunspots appeared so dark only by contrast to the exceedingly bright surface of the Sun; had it been possible to observe them in isolation, they would have been much brighter than the full Moon. Moreover, clusters of little planets orbiting at a relatively great distance from the Sun would have moved against the background of the latter at a practically constant velocity. Galileo had observed, on the contrary,

Fig. 2.4.1 - Laying out the "staves" for the skeleton construction.

Fig. 2.4.2 - Views of the "skeleton" structure.

that the motion of the sunspots was remarkably slower in proximity to the edge of the Sun than at the centre of its disc, in perfect compliance with the hypothesis of their contiguity to its surface.

As concerns the nature of sunspots, Galileo presumed that they consisted of vapour and exhalations generated by the combustion of solar matter; something comparable "to our clouds or fumes; and certainly anyone who wanted to imitate them with some of our own matter, could not, I believe, do so more easily than by placing on a red-hot iron plate a few drops of some bitumen hard to burn, which would impress on the iron a black spot, from which [...] would rise dark smoke that would expand and spread in extravagant, changing figures". Even more than the 'earthly' nature of the Moon, then, the fact that the Sun was subject to continuous and random changes profoundly undermined the whole edifice of Aristotelian cosmology. (*P.D.S.*)

2.4 - Re-creating Galileo's Telescope

Galileo's leather telescope with its gold tooling represents the type of instrument presented to customers and sponsors of high rank and great influence. It is a symbol of Galileo's keen marketing skills and brilliant entrepreneurship.

Re-creating this telescope was a challenging undertaking. In recognition of Galileo and all that he represents, the goal was to replicate the instrument as faithfully as possible, internally and externally. The process of researching the literature for what is known about the instrument, examining the original, and the actual reconstruction of the replica revealed many valuable insights. These included the technical expertise and state of the art construction techniques used by Galileo and others in the early 17th century.

Replicating the tube and lens holders: A study of the photographs taken during the 1996 restoration of the telescope revealed the internal wood construction throughout the instrument. Most notable is the wood (possibly bass or linden) that forms the tube: cut in strips configured in a most novel way. That is, the telescope consists of a skeleton structure of thin wood staves glued onto a backing cloth and formed into tubes similar to a barrel type construction. However, the edges were not chamfered and butted for radial strength as usually required to prevent these structures from collapsing – as found in early trials to replicate this technique. The tubular strength was achieved by liberal amounts of glue applied throughout the wood layers filling all the voids. Today we call this a resin composite construction in which the resin bonds all the materials together into a very light and strong structure with physical characteristics superior to the individual components. A wood skeleton tube made this way resists high crushing forces, enough to stand up to the strong pressures applied during gold stamping.

This was an enlightening insight into the inventiveness and advanced techniques of craftsmen in the 1600s. This technique was easily achieved with unskilled labourers and required only simple tools. The sheets of wood staves glued on cloth backing provided a common starting material easily shaped into the various telescope tube diameters, single and multilayered, including the main barrel, the objective and eye piece housings **(Figs. 2.4.1 – 2.4.2)**.

Replicating the artwork: While there is no evidence that Galileo himself would have done the leather gilding, his choice of the artisan was as critical to him as the rest of the details for producing this telescope. Precisely replicating the gold stamping designs along the entire length of the telescope with all their

Fig. 2.4.3 - Developing the artwork.

Fig. 2.4.4 - Embossing the gold bands on the curved surface of the finished tube.

variations and original irregularities was a major and necessary effort.

There is no documentation for any of the gilding patterns on the original telescope, so for this important step, the high resolution photographs were indispensable. Developing the artwork from photos was an exacting process and entailed complex issues. The shape and detail of each of the design elements were extracted from images that included all the dents and distortions from four hundred years of use and aging. These had to be differentiated from the normal and highly valued irregularities associated with any handcrafted artwork. Often two or more images had to be merged for a complete pattern to be developed. The computer images of the original gold tooling from cylindrical surfaces had to be 'flattened', scaled to the original size, and checked against the tube perimeters to ensure matching seams. The final black and white drawings also allowed for the usual image spreading during the stamping process **(Fig. 2.4.3)**. Deciphering all these details achieved a faithful match to the originals and resulted in more than twenty dies cast for the gilding process. These varied in size from a few millimetres to an average of two or three centimetres.

The gilding process: After the brown and the distinctive bands of red leather were applied over the wood staved tubes of the telescope, the final step of gilding was performed. The bands of leather give the telescope its quality and durability and the gold tooling gives it the elegance needed for a presentation piece. Inspecting the original indicates that the leather gild stamping was applied on the finished tube. Not many craftsmen would attempt this process today because it requires experience and skill. However, it was indispensable to learn and experiment with this process to properly replicate Galileo's telescope.

In the actual die stamping process the gold adhesive is usually the time honoured mixture of albumin (in the form of egg white) and vinegar. The metal stamp is heated to a temperature above the boiling point of water. The right temperature is determined by the age old method of touching the hot stamp to a moist sponge to check for the distinctive sizzle sound that signifies the perfect temperature has been reached. The gold particles are impressed into the grooves of the leather as the pattern on the hot stamp is pressed into its surface (**Fig. 2.4.4**). This process leaves a recessed image less vulnerable to wear and tear than surface images. More than four hundred individual hot stamps were applied on the overall replica.

The gold patterns on the original telescope were more pronounced when Galileo's telescope was new. However, they still speak eloquently of the importance of the instrument they adorn and to the impressive capability of craftsmen in the days when this type of skill was much more prevalent. (*J. & R.M.*).

Fig. 3.0.1 – Signed objective lens by G. Campani, 1655 (IMSS, inv. 2587)

Chapter 3:

The Quality of Galileo's Lenses

Pier Andrea Mandò, Luca Mercatelli, Giuseppe Molesini,
Maurizio Vannoni, Marco Verità

3.0 - Introduction

Although so fragile and modest, lenses have played a key role in laying the foundations for such a great part of our knowledge. Lenses were in use since Antiquity; however, it was only in early 17th century that they underwent a thorough process of quality development that resulted in almost perfect items supporting the advancement of astronomy and natural sciences. Quality is indeed of the utmost importance for a lens to perform properly; it certainly took a great effort to improve production technology, fine tuning all the steps of the craft to perfection. The development involved the glass production process, the grinding and polishing techniques, and the final tests for quality assessment. Successful artisans then started to sign their best lenses, as an artist would do with his masterpieces **(Fig. 3.0.1)**.

The history of this development, as well as that of the discoveries that it made possible, is written in the lenses themselves, embodied in their physical and geometrical characteristics. To read this information today, and reconstruct the history that is behind it, accurate scientific analysis with appropriate instruments and measuring techniques must be performed. The first step is individuating the lenses of significance still existing. Unfortunately, only a few pieces have survived the vicissitudes of time; the ones remaining are preserved in museums. The Institute and Museum of the History of Science in Florence houses a prominent collection of 17th century lenses; these have become an invaluable source of information for research aimed at clarifying the circumstances that enabled extraordinary celestial discoveries.

As for scientific analysis, due to the uniqueness of the pieces, only tests that are strictly non invasive can be carried out. While this in general restricts the range of possible investigations, it still allows functional tests, and the use of light as a probe. Modern measuring techniques and state-of-the-art optical in-

a

struments can provide accurate information on the geometrical shape of the lens refractive surfaces, which ideally ought to be spherical, or plane. Also the residual micro-roughness after polishing is of significance, since it is a record of the workshop technique used for manufacturing; detailed lens surface inspection can be made with microscopes and optical micro-topographers. Overall, the quality of lenses in relation to their geometrical features can then be assessed in a completely safe manner.

Although it appears obvious, the lens material – the glass – has to be homogeneous, free of tint and impurities, and transparent in visible light. Improving the glass quality in these respects was essential to the effective use of lenses in optical instruments. During the 17th century, special processes had likely to be developed to produce glass for optical applications. The specific composition of the glass is therefore important for determining the physical and chemical characteristics of the lenses, as well as for providing information about the development of glass production technology, and about the origin of the pieces under investigation.

The methods of analysis that can be used are optical spectrophotometry and X-ray fluorescence. The former relates to the glass transparency at different colours, and points out a few typical elements that may be present in the glass

b

Fig. 3.1.1 - Interference colours produced with soap films, a) in the form of a bubble, b) in wedged vertical layer.

melt. The latter is a more direct determination of the atomic elements within the glass, and is essential for identifying the glass components. A successful investigation, along with the expertise of glassmakers and the knowledge of optical technology historians, could lead to the deciphering of the production recipe and, eventually, the reproducing, in a faithful manner, the glass varieties used to make lenses in the 17th century. (*G.M.*)

3.1 - Optical Analysis of 17th Century Lenses

Lenses from the 17th century were studied for optical quality with state-of-the-art equipment, and with procedures currently used in optical metrology laboratories. Tests carried out were non-contact measurements based on interference of light.

Interference is a physical phenomenon due to the wave nature of light, showing up, for example, when looking at the colours of soap bubbles in sunlight **(Fig. 3.1.1a)**. Colours are produced by reflections at the inner and outer surfaces of the soap film, and encode the thickness of the water layer in-between. Making a vertical film with a frame, a sequence of colours is observed, accounting for a wedge-shaped water layer **(Fig. 3.1.1b)**.

Fig. 3.1.2 - Interference pattern produced with two almost matching glass discs, on top of one another.

In a similar manner, by placing two discs of glass on top of one another, between the almost contacting surfaces a thin layer of air is formed. The reflections from the upper and lower facing surfaces produce a sequence of colours, named "fringes", that encode the thickness of the air layer in-between **(Fig. 3.1.2)**. Naturally, if the lower surface is plane, the fringes map the shape of the upper surface. Similar results can be found if the facing surfaces are both spherical, one concave and the other convex, the fringe pattern revealing the departures in shape between them. By this basic mechanism, the shape of the lens refracting surfaces is studied.

Due to the polishing process, under uniform wear the expected shape of the lens surfaces is spherical. Departures from spherical shape are considered defects, producing a loss in quality in the end image. A common defect is uneven wear of the lens at the edges, resulting in some image blurring. Such a defect, as well as others that may arise, should be maintained below a threshold, not to affect the overall lens performance.

To reconstruct the actual shape of a lens surface, the interference pattern needs to be decoded. Modern instruments work on fringe patterns produced with laser light, and on-line processing with a personal computer. An example

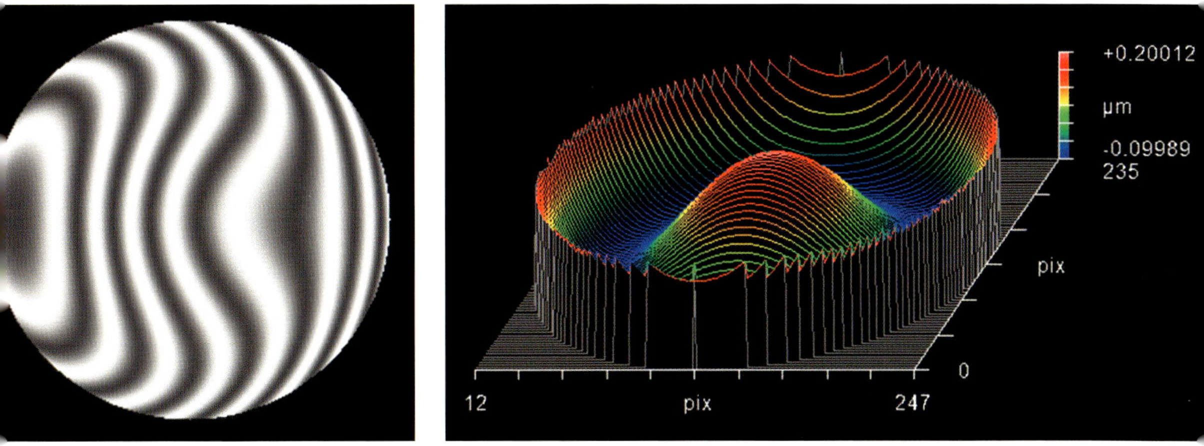

Fig. 3.1.3 - Example of a defective surface. Left: interference pattern. Right: reconstructed surface error.

is shown in **Fig. 3.1.3**, for a surface that is approximately 0.0003 millimetres from being a perfect sphere.

As for 17th century lenses, the telescope optics of Galileo, Torricelli, Divini and Campani housed at the Institute and Museum of the History of Science in Florence, have been examined during the last few years. The optical testing facility **(Fig. 3.1.4)** is located in a special laboratory that is called a class "C" ISO 7 clean room, with standard environmental conditions of 20 ± 1 °C temperature, and 45 ± 5 % relative humidity. The measuring instruments are placed on a granite table, passively isolated from the floor to reduce vibrations.

The data collected showed evidence of the enormous advances in optical technology that were accomplished at that time. In fact, at the dawn of the 17th century, the level of lenses was that of bare magnifiers. At the end of the century, they had become high-technology items that differ little from today's quality optics.

The key case is that of Galileo's lenses. He certainly realised that the performance of his telescope depended critically upon the quality of the lenses he could have (particularly, the objective). In the absence of a fabrication technique that could yet guarantee the end result, Galileo used to purchase great quantities of lenses, and then choose the ones that performed best. This process must have been very laborious and costly, also requiring the definition and establishment of appropriate testing criteria. Notwithstanding these difficulties, the quality of Galileo's telescope lenses was extraordinarily good. In addition to the proof given by the astronomical observations that he made, this conclusion is drawn after examining the original "broken objective lens" mounted in the ivory frame by Vittorio Crosten (1677). The diameter of the lens alone, free of its mounting (**Fig. 3.1.5**), is approximately 58 millimetres, but in use it was probably stopped to a somewhat smaller clear diameter with a diaphragm. The glass purity and homogeneity are still far from being perfect; several seeds are clearly visible. As for homogeneity, surface figure and overall optical performance, the lens is almost perfect for use in a telescope. This appears in the fringe pattern (**Fig. 3.1.6**), which relates to a clear diameter of approximately 38 millimetres; although the fringes have a different slope in the two main fragments (due to a residual mismatch of the parts after being glued together) their slight lack in straightness still attests to a performance close to ideal. The minor local irregularities of the fringes are attributed to small heterogeneity effects of the bulk glass.

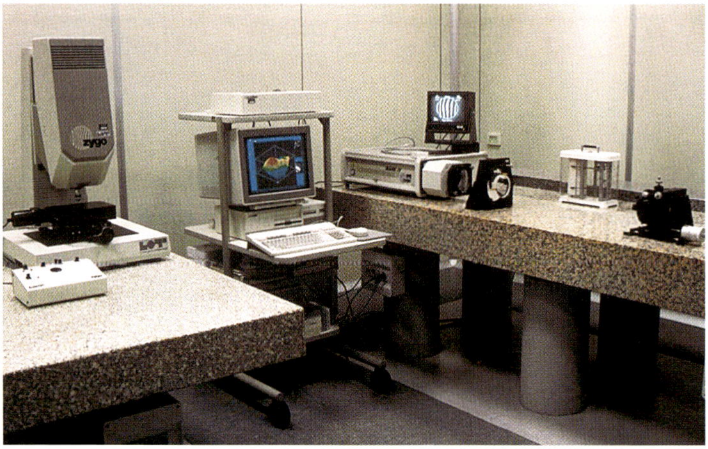

Fig. 3.1.4 - Optical testing facility where the lenses have been examined.

Fig. 3.1.5 - Galileo's broken objective lens, out of its frame (IMSS, inv. 2429).

Fig. 3.1.6 - Fringe pattern obtained after testing Galileo's broken objective lens. Straight and equidistant fringes mean high optical quality.

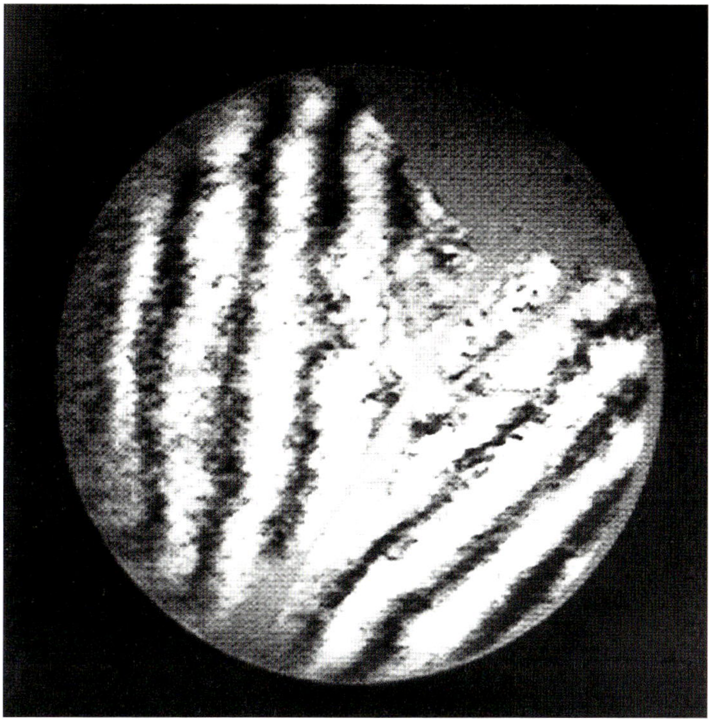

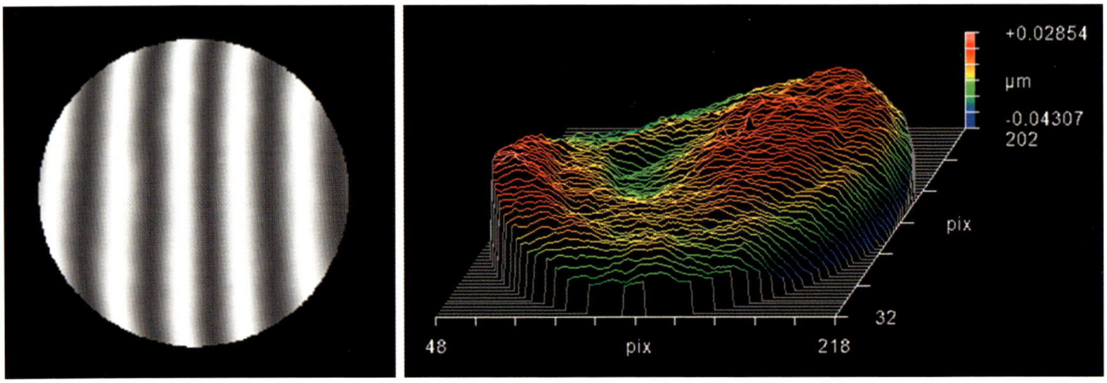

Fig. 3.1.7 - Interference pattern (left) and reconstructed surface error (right) of an eyepiece from a telescope by G. Campani, 1666 (IMSS, inv. 2556).

Fig. 3.1.8 - Terrestrial telescope with six tubes by G. Campani, 1666 (IMSS, inv. 2556).

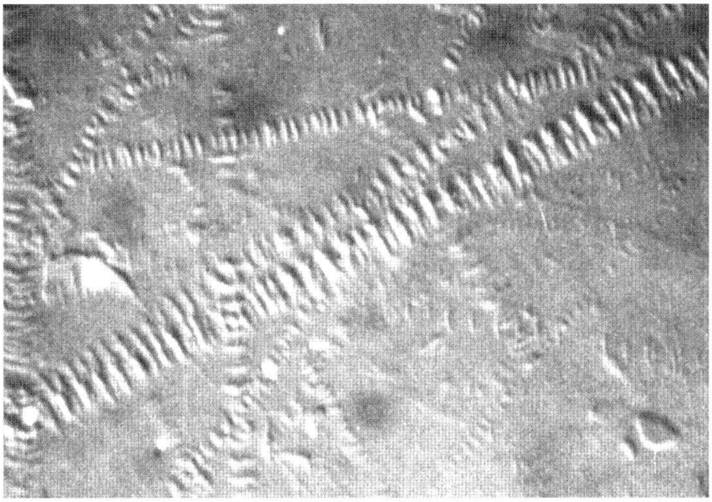

Fig. 3.1.9 - Sample Nomarski micrograph of a lens surface polished by Campani. The area displayed is 0.20 x 0.15 square millimetres.

It is understood that, as the century progressed, the lens polishing mastery of the artisans and the introduction of appropriate tools and techniques reduced the discarding of imperfect pieces and allowed for the producing – in quite a reliable manner – of good quality lenses. A testimony to this achieved ability to polish lens surfaces to an almost perfect shape is presented in **Fig. 3.1.7**. The lens being examined belongs to a telescope made by Campani, in Rome, in 1666 **(Fig. 3.1.8)**. The residual surface imperfections are extremely small, and produce negligible effects on the image quality.

Information on the polishing process of lens surfaces in the optical shop can be obtained from the observation of the surface micro-topography with a microscope. Ideally, no features should be visible. In practice, however, small digs, pits, scratches, stains, dust particles and all kinds of contaminations appear. Such defects, partly due to glass weathering and lens handling, tend to scatter light and produce a background haze in the image that is being observed. To detect the surface structure, a differential interference contrast microscope of the type invented by Jerzy Nomarski (1919-1997) – and for this reason briefly called a Nomarski microscope – is often used, since it enhances the surface relief as if it were illuminated sideways.

One of such Nomarski micrographs obtained from a lens surface polished by Campani shows, among other features, evidence of major scratches resembling tire treads in sand **(Fig. 3.1.9)**. The latter, which are observed in some earlier optics, are attributed to the rolling grains of abrasives used during the lens polishing process. The surface finish shows that Campani used a sort of gentle grinding technique, avoiding excessive pressure between lap and glass. This might have been one of his craft secrets that allowed him to produce very thin lenses without compromising the excellence of the surface form.

For a more quantitative analysis of the surface features, non-contact 3D optical profilometry with a laser or a white light micro-interferometer can be used. In **Fig. 3.1.10** the height map of a small area of a lens surface polished by Divini is presented. The spherical curvature of the surface is subtracted via software. Again, scratches and various defects are visible. To the eye of the expert, the major spots that appear account for a polishing technique heavily relying on pressure between lap and glass, leading to a better finish of the surface in terms of microroughness, however, at the expense of the overall quality of the surface form.

In addition to the regularity of the surfaces, the lens performance depends on the quality of its constituent bulk material: the glass. In the absence of specific melts for use in optics, in the early 17th century the artisans used to make lenses selecting the glass for purity, transparency, homogeneity and absence of any tint (colourless glass). Studying the actual composition is useful in identifying the region of provenance of the lens material, and the process of glass production as well. Information can be derived both from optical spectrometry and elemental analysis.

Optical spectrometry consists of a photometric measurement of glass transmittance, varying the colour of the illuminating light. In a schematic device to accomplish this task **(Fig. 3.1.11)**, the light emitted from a lamp is cast on a screen with an opening in the form of a slit, to produce a narrow beam of white light that impinges on a prism. The prism separates the colours of the beam, so that a second slit on a screen selects the actual colour that is passed (e.g. green). The latter is made to go through the glass sample under test, and is partially absorbed according to the features of the glass. Finally, the transmitted light is detected and processed, finding out the amount of light which is absorbed at that colour. Ro-

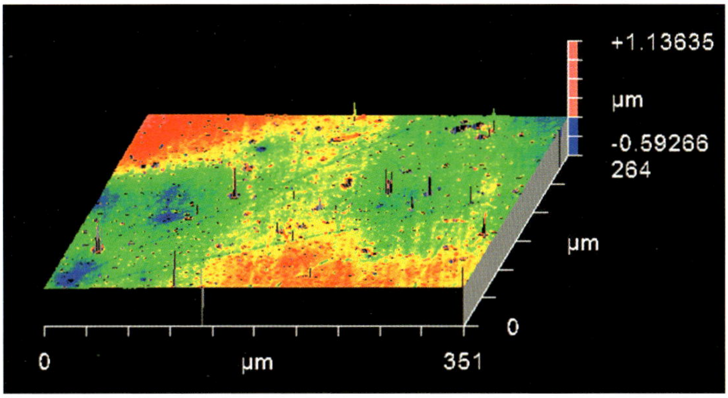

Fig. 3.1.10 - Sample micro-interferometric map of a lens surface polished by E. Divini. The area displayed is 0.35 x 0.26 square millimetres.

Fig. 3.1.11 - Schematic layout of a device measuring the spectral transmittance of a sample lens.

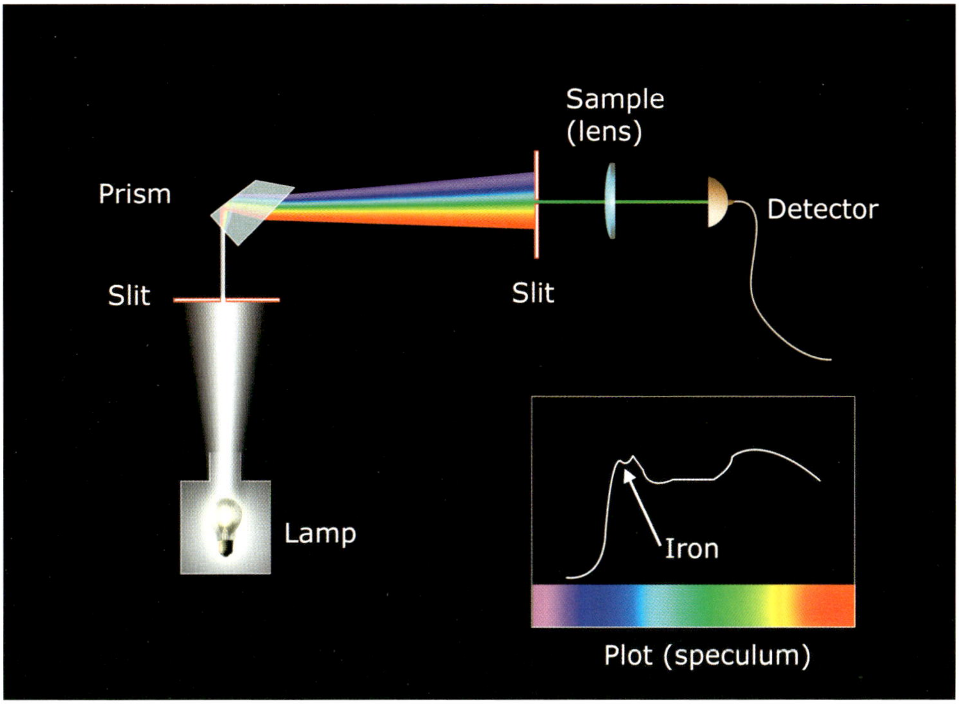

tating the prism displaces the spectrum across the second slit, so that a different colour is selected for measurement. The operation is repeated for all the colours of the spectrum. By collecting all the measurements in a single plot, a curve of transmittance as a function of the colour is obtained. The curve exhibits peaks and dips that are typical for particular elements, and are useful in identifying the ingredients present in the glass melt.

Features of the spectrophotometric transmittance curves for the lenses made by Divini, Torricelli, Mariani and Campani **(Fig. 3.1.12)**, actually point out the presence of iron (a common glass contaminant), plus a certain amount of manganese. Such an ingredient was known to be effective in eliminating the glass' tint due to iron (discolouration of green to yellow and blue light colour).

Fig. 3.1.12 - Transmittance curves of 17th century lenses housed at the Institute and Museum of the History of Science in Florence.

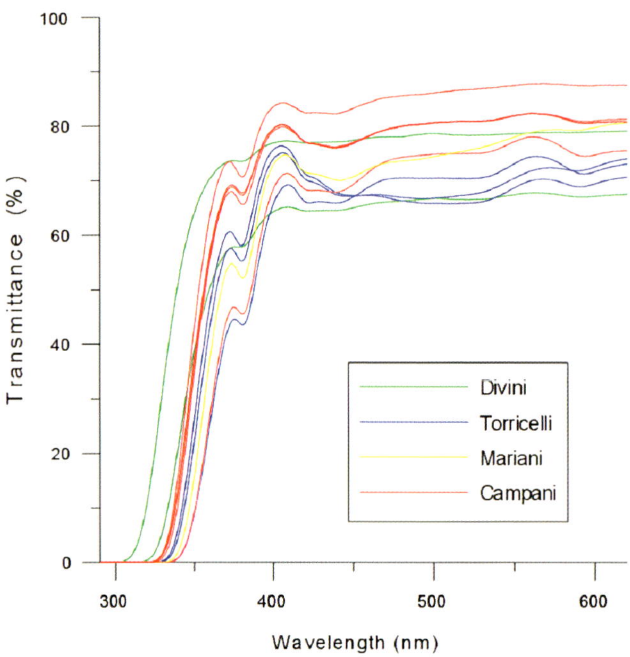

Of course, for proper evaluation, all the above information needs to be complemented with analysis performed by elemental analysis, and with knowledge from different fields of science, including studies on the aspects of early technologies, in order to reconstruct the entire identity of the lenses and the optical instruments under study. (*G.M., M.Va., L.M.*)

3.2 – Analysis of Lenses with the XRF Technique

With the so-called XRF (X-Ray Fluorescence) technique, the composition of a material can be analysed in totally non-invasive manner. Utilizing portable instrumentation, this technique can be used to perform on-site analyses, in museums for example.

The XRF technique is one of the so-called 'elemental' techniques; that is, it recognizes the chemical elements present in a material, but does not distinguish the type of molecule in which these elements are bound. In very schematic terms, during the course of an XRF measurement:

• a flow of primary ionizing radiation is sent to the material to be analysed;

• the material 'reacts' by emitting X-rays that have different energy levels depending on which atoms are present in the material;

• a special instrument detects the X-rays emitted and recognizes their different energy levels, reconstructing what is called 'the energy spectrum' of the X-rays pertinent to the sample;

• by examining this spectrum, we can deduce which elements are present in the zone subjected to ionizing radiation.

More specifically, the flow of primary ionizing radiation can be produced by an X-ray tube, similar to those used for radiographs. However, in the case of XRF, the tube can be of extremely reduced size and weight (a few centimetres, a few hundredth grams), which makes the instrumentation very easy to transport. In alternative to an X-ray tube, the ionizing radiation can be emitted by another type of radioactive source that emits either X-rays or gamma rays.

Through the so-called 'photoelectric effect', the interactions of the X-rays or gamma rays with the atoms in the 'target' result in ionization of the innermost orbitals (the regions of probability in which the electrons are found) of the atoms. The electrons that are closest to the atomic nucleus may be displaced from their original positions, leaving a 'gap' in the related orbital **(Fig. 3.2.1a)**. Instantaneously, an electron occupying a position further away from the nucleus makes

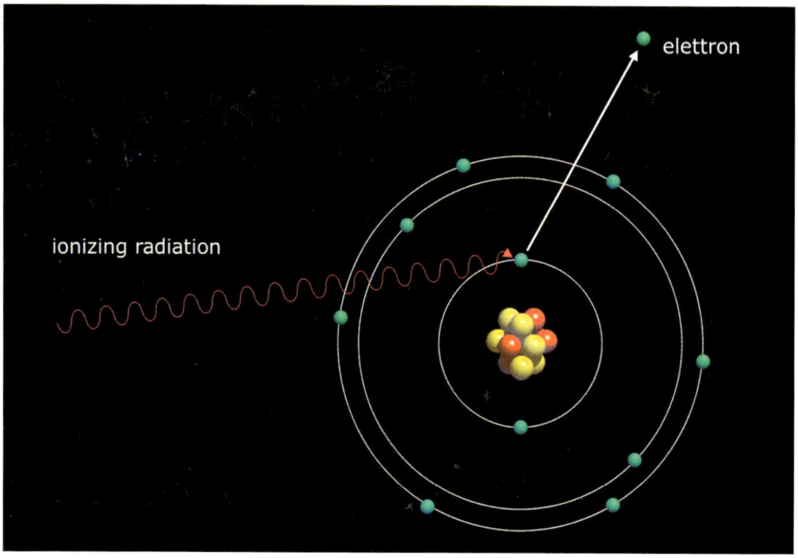

Fig. 3.2.1 - Above: Ionizing of an electron by photoelectric effect; Below: the phenomenon of X ray fluorescence.

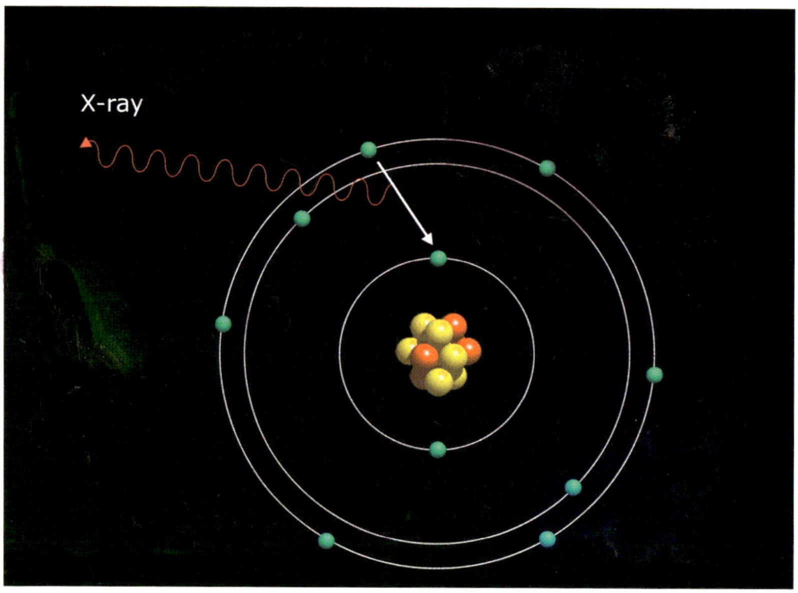

a transition, going to fill the gap that has formed. Since the energy E_I possessed by this electron in the more distant position is greater than the energy E_2 possessed by it after the transition, the available energy ΔE, equal to $E_I - E_2$, is 'emitted' in the form of an X-ray, having energy ΔE **(Fig. 3.2.1b)**. Since the energy levels of the electrons in the different orbitals are characteristic for each given chemical element (for example: potassium, calcium, iron, lead, etc.), the energy levels ΔE of the X-rays emitted are also characteristic of the element. For each element, the energy levels of the characteristic X-rays are well known. If then we have available a detector that recognizes the energy levels emitted during measurement, we can determine which elements are present in the area subjected to the flow of primary ionizing radiation.

In a single measurement, the elements that are predominant, minor, or present only as traces can be detected simultaneously. The minimum detectable concentration (that is, the 'sensitivity' of measurement) depends on the atomic number of the element. However, sensitivity to the different elements also varies depending on the measurement conditions; but in general it may be considered

Fig. 3.2.3 - The energy spectrum of Galileo's broken objective lens.

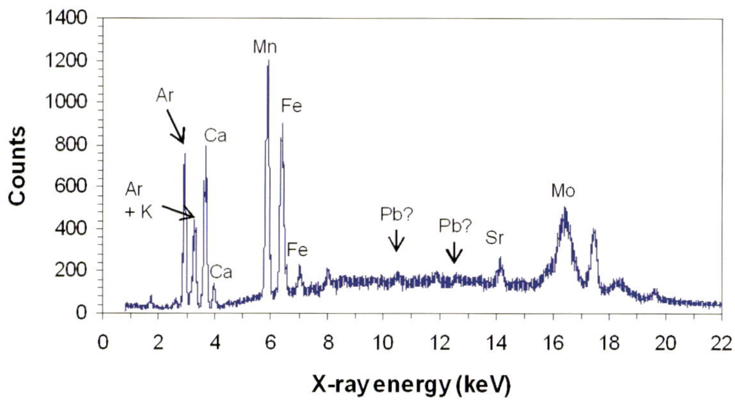

very good for the elements with high atomic numbers, while it is low or practically null for the light elements. Specifically, with the portable measurement systems, detection is good only for the elements starting from sulphur (atomic number, 16), while the lighter ones such as carbon, oxygen and sodium cannot be detected. Aluminium and silicon are also hard to detect, except when present in very large quantities.

With the portable versions of the XRF instrumentation, the results obtained remain at a semi-quantitative level. It is however possible to compare the quantities of the different elements present in materials of similar composition, such as different types of glass.

The results of the first exploratory campaign of tests conducted on some lenses from instruments constructed by Galileo, Divini and Campani do not yet allow us to draw any general conclusions that may be considered statistically significant. However, the lenses of Divini and Campani subjected to analysis appear to be substantially similar in composition, although with minor differences between one lens and another as regards the content of manganese, which is however present in a minor amount. Moreover, none of the lenses made by Divini or Campani was found to contain lead except in minimum traces, of no significance.

Galileo's broken objective lens appears instead to be made of glass of notably different composition **(Fig. 3.2.2)**. It too, however, contains no lead except in traces. Obviously, until measurement has been extended to a significant sample of the various scientists' production, it will be impossible to know whether this difference may be considered characteristic, or whether it merely pertains to the specific lens analysed. It will then be necessary to evaluate whether, and to what degree, the different composition, should it be found systematic, was purposeful or merely contingent, and whether it may have influenced the quality of the astronomical observations. (*P.A.M.*)

3.3 - European Glassmaking in Galileo's Time: Glass Composition and Quality

A significant number of analyses of Renaissance glass remains are available today. Along with study of the Medieval and Renaissance treatises on glassmaking, these analyses allow for a good understanding of the evolution of glass composition and technology from different glassmaking areas throughout Europe. Moreover, an early 17th century recipe book by a Venetian glassmaker called Darduin and a small number of documents on glassmaking from northern Europe have been published in the last twenty years.

Remarkably, the first printed book on glassmaking, *L'Arte Vetraria* (1612) by Antonio Neri, was known to Galileo. On January 3rd, 1614, Federico Cesi asked him for a copy of the book, and Galileo promptly sent it to Rome. Our knowledge of Neri's history is only fragmentary. He is reported as a priest and chemist. He developed his knowledge of glassmaking in Pisa and Florence (both centres strongly influenced by the Venetian glassmaking tradition) and later in the Netherlands and Antwerp.

Glass composition: Each of these texts describes the use of two basic raw materials: sand or quartz pebbles as a silica source, and plant ash as a fluxer. Despite this apparent simplicity, the glassmakers had to face several problems arising from the complex and varying composition of the natural raw materials.

Since the mid-14th century, Venetian glassmakers had obtained high quality products by replacing sand with quartz pebbles, which testifies to the efforts made to find a high purity silica source. The pebbles came from the rivers Ticino and Adige (those coming from the latter were inferior in quality). To reduce them to a finely ground powder suitable for melting, they were roasted, cast into water, and subsequently ground and sieved.

In Italy the main glassmaking centres were Venice, Tuscany and Altare. In the Mediterranean area, ash of coastal plants (*Salsola kali*, Salicornia, etc.) was used, while in northern-central Europe ash of inland plants (fern, beech, oak, etc.) was used. However, the Muranese glasshouses were expressly forbidden to make use of wood ash. Important differences in composition exist between coastal and inland plant ashes. Coastal plant ash is made mainly of sodium and calcium carbonates, but also of chlorides, sulphates, phosphates and small amounts of potassium, magnesium, iron, aluminium and silica. Instead, the ash from inland plants (wood ash) consists mainly of potassium and calcium carbonates, sulphates and phosphates and small amounts of other elements. Chlorine is almost absent, but traces of manganese (MnO up to 0.5%), absent in the coastal plant ash, are present. Furthermore, important compositional variations exist between ashes of inland plants, depending on the plant species, provenance (same species growing in different locations), plant parts, burning temperatures and methods, etc. Less pronounced differences are reported for coastal plant ash.

Fig. 3.3.1 – Calcar to produce frit (G. Agricola, *De re metallica*, 1556).

The glass technology of this time ignored the use of lime (and magnesia) as stabilizing oxide. Despite its importance for the resistance of glass to weathering, it was introduced accidentally into the composition in the form of ash component. At the moment, it is not known exactly where and when (probably in Bohemia during the 17th century) the voluntary addition of a source of lime became a common procedure.

In Venice, ash was imported from the Levant, purchased in Syria or in Egypt (the Egyptian product was probably not used for clear glass, for it was thought to be less pure). Its trade was protected by the Venetian government, with a view to restricting its use. The astonishing consistency of composition for the glass manufactured at Murano over several centuries, demonstrates the importance of this raw material.

In central-northern Europe a wood ash glass had replaced the traditional natron glass since the Carolingian period. The production of this potash-lime-silica glass continued until the 18th century. Starting in the 15th century, spruce ash was also used and a new composition called wood-ash-lime glass appeared. The amounts of calcium oxide (CaO) and potassium oxide (K_2O) vary greatly in these types of glass, forming an almost continuous series of compositions between high potash and high lime glass.

A third raw material, manganese oxide (pyrolusite), was the decolorizing agent used to neutralise the light green to yellow tint deriving from iron impurities in the raw materials. In Venetian glass furnaces, manganese was added directly to the melt in controlled amounts, until decolouration was obtained. In this way the use of manganese was limited to the lowest quantity necessary. Until the end of the 16th century, the decolouration of glass in northern-central Europe was problematic. In this sense, the presence of manganese in the inland plant ash was a drawback as compared to coastal plant ash (practically free from it). During the 17th century (it is not known where and when) manganese was replaced by the more efficient arsenic.

The procedure for making glass consisted of fusing silica and plant ash in known proportions. The melting technique did not permit very high temperatures to be attained in the furnaces. Hence, the melting process was carried out in two steps. In a first step the ash and the sand were raked together in a reverberatory furnace (the calcar) at relatively low temperature for several hours, causing the material to sinter and avoiding the formation of a vitreous phase **(Fig.**

3.3.1). The product thus obtained, called frit, was then transferred to a pot furnace where it was melted at higher temperature **(Fig. 3.3.2)**. This second stage could last from twelve hours to several days. A third furnace, or the cooler part of the former furnace, was required to cool products from 500 °C to room temperature to relieve stress (annealing).

Glass quality: No special glass for lenses was made in Galileo's time and the optical quality of the glass produced was affected by several defects. It was thus impossible to find a piece of glass completely free of seeds, cords, a certain turbidity and slight colour (or grey). In 1610, Galileo succeeded in persuading Cosimo II to have a special glass expressly for lenses made in Florence. This experiment was probably not successful, for in October 1618 Galileo asked Sagredo to assess whether in Murano "the glass for making lenses can be improved".

The origin of these defects arises from the technological limits of glassmaking at that time. The temperature reached in the melting furnace was too low to allow complete refining (elimination of gas bubbles and seeds) and ensure good glass homogeneity. Layers (cords) or knots of glass with a different refraction index inducing light distortion can frequently be observed in ancient glass. Furthermore, the glassmakers would be faced by the fact that the considerable amounts of insoluble salts (sulphates and chlorides) of the plant ash are practically not reactive with the silica and could not be incorporated into the glass. If not eliminated, such salts cause a dispersion of light resulting in glass turbidity.

Finally, the quality of the raw materials was not as high as today and colouring impurities (mainly iron minerals) were introduced into the glass. In Galileo's time, the iron concentration in clear glass was high (Fe_2O_3 between 0.2 and 0.6%) as compared to modern optical glass (less that 0.01%). The iron content and its oxidation state produced a more or less intense hue ranging between yellow, green and blue. To obtain a colourless glass, the hue was eliminated by adding manganese oxide, but this process of glass discoloration led to the production of a grey glass.

Thanks to the skill and tradition of glassmakers, these defects were reduced, but not entirely eliminated. For instance, Venetian glassmakers used the technique of casting molten glass into water and subsequently re-melting it. This procedure, repeated several times, together with the scumming of the molten glass, helped to reduce un-reacted chlorides and sulphates, improving the clar-

Fig. 3.3.2 – Melting furnace (G. Agricola, *De re metallica*, 1556).

A
B
B
B
C
C
C
A
A
A
E
A
A
D
E

ity of the glass. From the Middle Ages onward, Venice was the most important glassmaking centre in Europe. Glass of good quality had been available there since the end of the 13th century, when glass lenses were first used in Europe. Furthermore, since the 15th century, the grinding and polishing techniques used to make mirrors had reached a high state of development.

Starting in the mid-15th century, Venetian documents classified clear glass into three types: "*vetro comune*" (common glass, slightly coloured), "*vitrum blanchum*" (uncoloured glass) and "*cristallo*" (crystal glass). At the time, the *cristallo*, invented by the Muranese Angelo Barovier (1405-1460), had acquired such clarity and homogeneity as to be compared to natural rock crystal (quartz). Soon it was traded throughout the world, and the secret formula was one of the main factors that allowed Venice to maintain its predominance over other European glassmaking sites for nearly two centuries. For the preparation of *cristallo*, Barovier added a preliminary step to the existing technique, consisting of purifying the plant ash. The raw ash was ground, sieved, and dissolved in boiling water. The resulting solution was filtered, concentrated and dried. The salt obtained was mixed in the right proportion with silica to prepare the frit. The purification process lead to the elimination of insoluble iron compounds (colouring impurities) as well as calcium and magnesium compounds, which are essential to stabilize glass against weathering. The glass obtained thereby (practically sodium silicate) would become coated with an opaque, weathered layer even shortly after its production, with consequent loss of the optical properties. Murano glassmakers searched for and discovered a solution to the problem. The source of lime and magnesia probably used was the *vitrum blanchum* frit or cullet added in the minimum amount necessary to stabilize the glass. This would cause only a moderate loss in transparency and clarity, while still yielding a product of higher quality than the clear glass known up to that time. Chemical analyses of Venetian clear glass from the 16th and 17th centuries detect in the crystal glass a lower calcium oxide (CaO) and magnesium oxide (MgO) content (as well as Fe_2O_3) than that of common glass and *vitrum blanchum*.

Several reasons caused the decline of Venetian glassmaking in the 17th century, in particular the invention of new kinds of glass in Bohemia and England. In Bohemia, where the glass industry was well established at this time, potash was obtained by lixiviation of the wood ash following the process invented in Venice. Stabilized by the addition of a source of lime (chalk), potash-lime-silica glass probably became clearer and brighter (with a higher index of refraction) than the Venetian *cristallo*. Lead glass (lead-silica glass) and plant ash-lead glass (potash-lime-lead and soda-lime-lead glass) had been used for naturally green or yellow coloured beakers and window panes since the 12th-13th centuries. But only in the latter part of the 17th century (1675 is the 'official' date) was the process improved upon by George Ravenscroft (1632-1681) in England, and the new lead-crystal glass, more transparent and with high refraction index, became available on the market.

In conclusion, at the time when Galileo was developing the first telescope, several kinds of glass of good quality were available not only in Venice, but also in Europe. Therefore, analysis of the lens composition of Galileo's telescopes will help to provide information on the nature, technology and quality of the glass used and to identify its provenance. (*M. Ve.*)

Chapter 4:

Observing with Galileo's Telescope

Paolo Del Santo, Simone Esposito,
Giorgio Strano, Maurizio Vannoni

4.0 - Introduction

Reconstructing events from the past is always an arduous endeavour, in which the historian must make use of all possible sources that may be of aid. Moreover, historiographic sources are sometimes contradictory, making it necessary to investigate sources of various kinds in order to corroborate or, vice versa, to refute, a working hypothesis.

In addition to the traditional sources (manuscripts or printed texts, laboratory notes, iconography, etc.), scholars of the history of science have at their disposal other material sources, as well as an interpretative approach of a very particular kind in the sphere of the historiographic disciplines. In cases where a reliable or sufficiently detailed description of an instrument or experimental apparatus exists, or better still, where the instrument or apparatus has survived intact and functioning to our own day, the historian of science can use it again to replicate experiments or observations made in the past. This methodology, known as "experimental history of science", has cast light on some scientific events of dubious authenticity; for example, in determining whether a scientist really did make a certain observation or experimental measurement, or only claimed to have done so.

There is of course no shadow of a doubt that Galileo assiduously conducted astronomical observations using the telescope. In this case, an approach of the experimental type can however furnish useful information on the performance of the first telescopes, not directly obtainable theoretically in accordance with geometrical optics. Certainly, there is no vision without interpretation, but just for this reason it is important to try to distinguish precisely between what Galileo was potentially able to see "objectively" with the telescope and what was, instead, the product of the interwoven physiological, psychological and cultural – in a word, interpretative – factors that assigned meaning to his observations **(Fig. 4.0.1)**. It is with this aim that the Institute and Museum of the History of Science

Fig. 4.0.1 - G. Galilei, *Opere*, Bologna, 1656: Frontispiece.

and the Arcetri Astrophysics Observatory have launched a research project designed to materially retrace Galileo's stages of observation.

At the moment, this project is still in the stage of perfecting a suitable experimental device: an optical replica as faithful as possible to the original instrument, fabricated on the basis of analyses conducted on Galileo's telescopes by the National Institute of Applied Optics and the National Institute of Nuclear Physics in Florence. But when all of the details have been completed, the experimental device will have the same characteristics as the refined telescope with magnification of about 20 times donated by Galileo to Grand Duke Cosimo II.

The images obtained with this experimental device, captured through modern photographic techniques, will record as faithfully as possible what the human eye would see through Galileo's telescope. Although these images will not solve historiographic problems, their explanatory and educational value will be exceptionally high. As has emerged from the earliest results achieved in the various stages of developing the device, the images will present those who examine them with the same problems of reading and deciphering the visual information experienced by the first users of the astronomical telescope (*P.D.S.*)

4.1 – Replicating Galileo's Observations

To replicate Galileo's observations, an optical device faithful to his original telescope, but also able to support suitable instruments for recording the images, has been developed. Like the telescope, this device is composed of two lenses: a converging one, called objective, and a diverging one, called eyepiece. When the eye is approached to the diverging lens, it sees an enlarged image of the distant object being observed. The lenses utilised are replicas of the original lenses in Galileo's telescope, constructed on the basis of detailed measurements of the shape and refraction index of the glass performed by the National Institute of Applied Optics. The mechanical structure that supports the lenses and allows the distance between them to be regulated was produced in the workshops of the Arcetri Astrophysics Observatory. The distances between the lenses have been calculated to provide images with the same field of view as that of the instrument used by Galileo. Lastly, the objective in this device has an entrance diaphragm the same size (about 15 mm) as that of the original telescope.

For recording the images, in place of the eye or a photographic plate, a modern digital device called Charge Coupled Device, abbreviated as CCD, has been

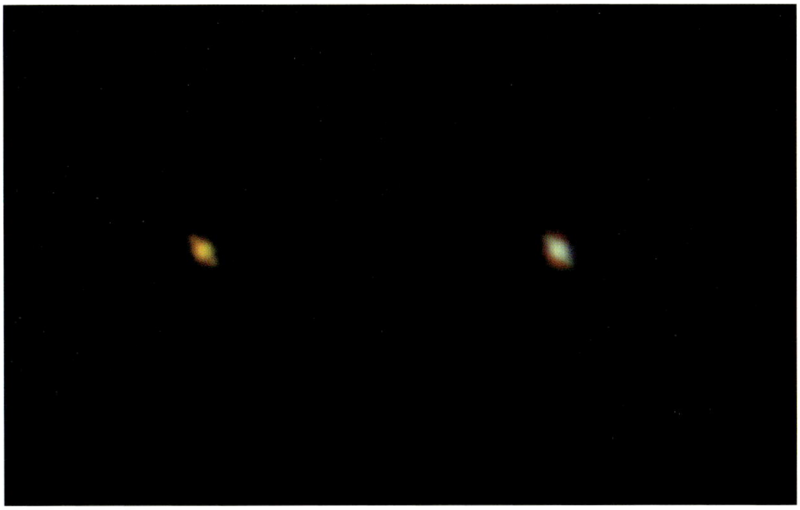

Fig. 4.1.1 - Saturn observed through the optical replica of Galileo's telescope. At left, image with yellow filter; at right, without filter.

used. The device, which operates according to the so-called "photoelectric effect", transforms the photons striking its surface into electrons that are detected and measured by special electronic control instrumentation. With the CCD, the photons that fall on each of the individual elements sensitive to the instrument, called pixels, can be recorded. The photographic CCD utilised in replicating Galileo's observations is composed of 2300 x 3400 pixels, corresponding to an image defined as a rectangular grid of 2300 x 3400 points. The digital image is a faithful copy of the one that would form on the retina of a human eye placed behind the telescope's eyepiece.

The most important characteristic of the images obtained with a telescope is their resolution. This quantity is defined by the minimum size of an object distinguishable in the image, and determines which structures are recognisable. For example, if the image of a building has a resolution of one meter, the windows and other parts smaller than one meter in size will not be clearly visible. In the case of Galileo's telescope, the resolution is worsened by chromatic aberration, due to the fact that the glass of which the lenses are made produces slightly different effects for light of different colours. In practice, the focal distance of a lens depends on the colour – that is to say, the wavelength – of the light that is being observed, so that it is impossible to bring all of the wavelengths observed into focus simultaneously. Consequently, the image appears wider than its real size.

Recording the image of Saturn by means of CCD provides a good example of the significance of chromatic aberration in Galileo's telescopic observations **(Fig. 4.1.1)**. As compared to CCD recording, it is however interesting to note that the human eye detects radiation only in a small fraction of the visible spectrum, a fact that limits the extent of the chromatic effect. A second example of the consequences of this aberration emerges in observing images of the Moon obtained with and without a coloured filter. In this case the aberration takes the form of a blue fringe on the right side of the image **(Fig. 4.1.2)**.

The astronomical images obtained in this way do not, however, take account of possible defects in eyesight other than myopia. While myopia could be compensated for by adjusting the relevant position of the objective and eyepiece, other visual defects, such as astigmatism, could not be corrected. Consequently, there still remains an aura of mystery as to how the Moon, Saturn, and Jupiter's satellites appeared to Galileo, the precise physical characteristics of his eyes being unknown. (*S.E.*)

Fig. 4.1.2 - The Moon observed through the optical replica of Galileo's telescope. At left, image with yellow filter; at right, without filter.

4.2 - Analysing Early Telescopic Observations

In addition to replicated observations, it is possible to esteem the quality of early telescopes – Galileo's in particular – by following another method. This requires: 1) the acquisition of good quality contemporary images of the celestial bodies, 2) the creation of a computer software capable of degrading the original quality of such images, and 3) the comparing of the degraded images to the early 17th century iconographic sources.

In order to establish the quality of an image, and to what degree a good quality image has been degraded, it is important to fix a quantity to measure, the so-called 'resolution'. In astronomy, resolution is the minimal angle below which a detail of a celestial body can no longer be distinguished. Resolution is measured in arc-seconds; 60 arc-seconds (or 60") forming one arc-minute, and 60 arc-minutes (or 60') forming one degree. To understand what one arc-second is, just take a one euro coin (23 millimetres in diameter) and place it at a distance of about 4.5 kilometres away from you. The angle that the diameter of the coin will subtend to the eye of the observer will be 1". Of course, it would be impossible to see the coin placed at such a distance by the naked eye. The scientific reason for this is that the optical device called 'the human eye' has a poor resolution power of about 120". In order to distinguish the coin, the distance must be decreased to approximately 40 meters. However, at such a distance, it would be impossible to see any of the coin's details, because its diameter will subtend an angle that is equal to the resolution power of the eye. To be able to distinguish details, it is necessary to reduce the distance of the coin even more... or to use a telescope!

Essentially, the software used to degrade good digital images, takes the mean value over all the pixels (the 'picture elements' of a digital image) within a circle of a given diameter, and use it to produce the pixels for a new image. This process produces a loss of visual information that is perceived as a blurring, which has a direct relation to the resolution. The two celestial bodies analysed by this method, the Moon and Saturn, have mean apparent diameters of about 31' (or 1,860"), and about 39" (rings included) respectively. Therefore, for a good digital image where Saturn appears 390 pixels wide (rings included), 1" corresponds to 10 pixels. The software will produce a new image with a resolution of 1" by taking the mean value over circles with a diameter of 10 pixels.

The comparison of degraded digital images to early representations of the planets is very intriguing, and answers many questions about early telescopic

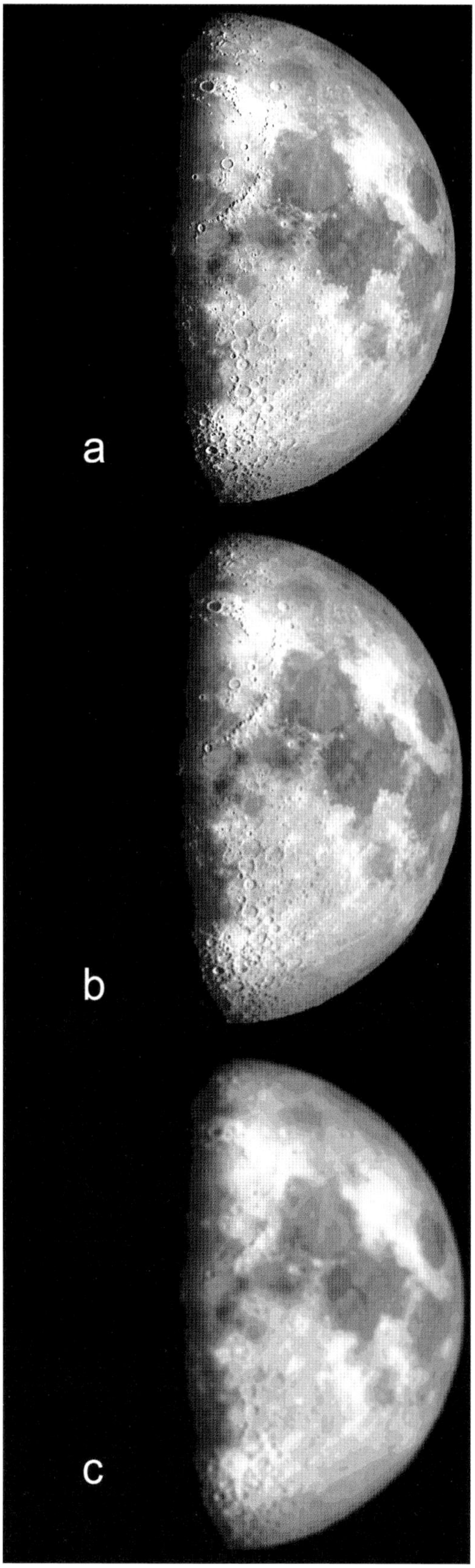
a
b
c

observations. By varying the resolution, a considerable number of phenomena observed by Galileo and his followers can be reproduced. Since the resolution power does not depend on the magnification, but on the optical quality of the telescope – in particular on the size and perfection of the objective lens – the features observed of the Moon and the planets are main indicators of the ability of early opticians.

The Moon: Images of the Moon degraded up to 10" and 20" of resolution **(Fig. 4.2.1)** reveal details far beyond those perceivable in early pictorial representations by Galileo **(Fig. 4.2.2)**. However, the engravings included in *Sidereus Nuncius* **(Fig. 4.2.3)** show details that suggest a resolution limit of about 10". Indeed, lunar craters are carefully delineated within the booklet. This simple comparison confirms Galileo's claim that the quality of his telescopes was considerably increased between late summer and late autumn of 1609.

Fig. 4.2.1 - Series of degraded images of the Moon one day after the first quarter at different resolutions: a) original, b) *c.* 10", c) *c.* 20".

Fig. 4.2.2 - Galileo's watercolour of the Moon a couple of days before the first quarter (BNCF, Ms. Gal. 48, c. 28, detail).

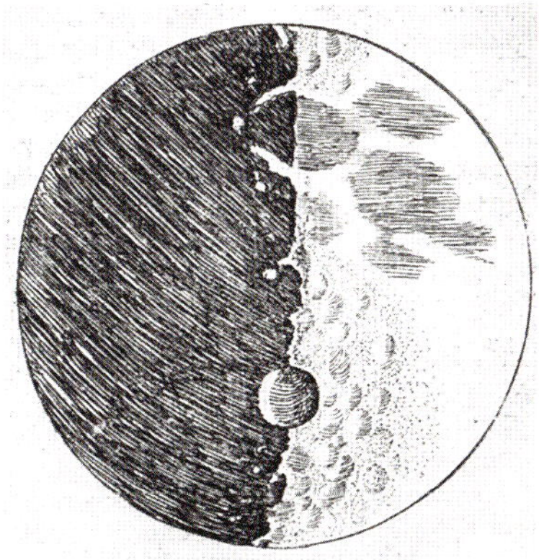

Fig. 4.2.3 - Engraving of the Moon at the first quarter (G. Galilei, *Sidereus Nuncius*, 1610, p. 9v).

This comparison bring to light two additional characteristics of the images in *Sidereus Nuncius*. Firstly, it is impossible to identify the outstanding crater that appears near the centre of the Moon during the first and the last quarters. With high probability, this crater is a mere example of the lunar relief as perceived by Galileo. Therefore, the engravings of the booklet are the result of the compounding of real telescopic observations with emphasised details of the lunar surface for explanatory purposes. These details depend on careful observations of the lunar relief that Galileo accomplished with the improved upon telescope he also used to discover the satellites of Jupiter.

Secondly, it is apparent that Galileo's engraving of the Moon at the first quarter does not include a lunar sea, more exactly the *Mare Serenitatis*. Probably,

Fig. 4.2.4 - Series of degraded images for Saturn, with the ring less or more inclined to the observer, at different resolutions: a) original, b) *c.* 2", c) *c.* 5", d) *c.* 10".

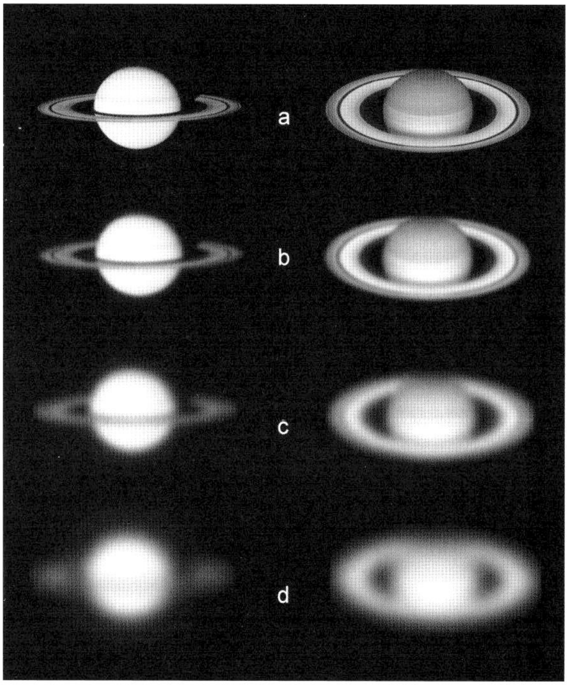

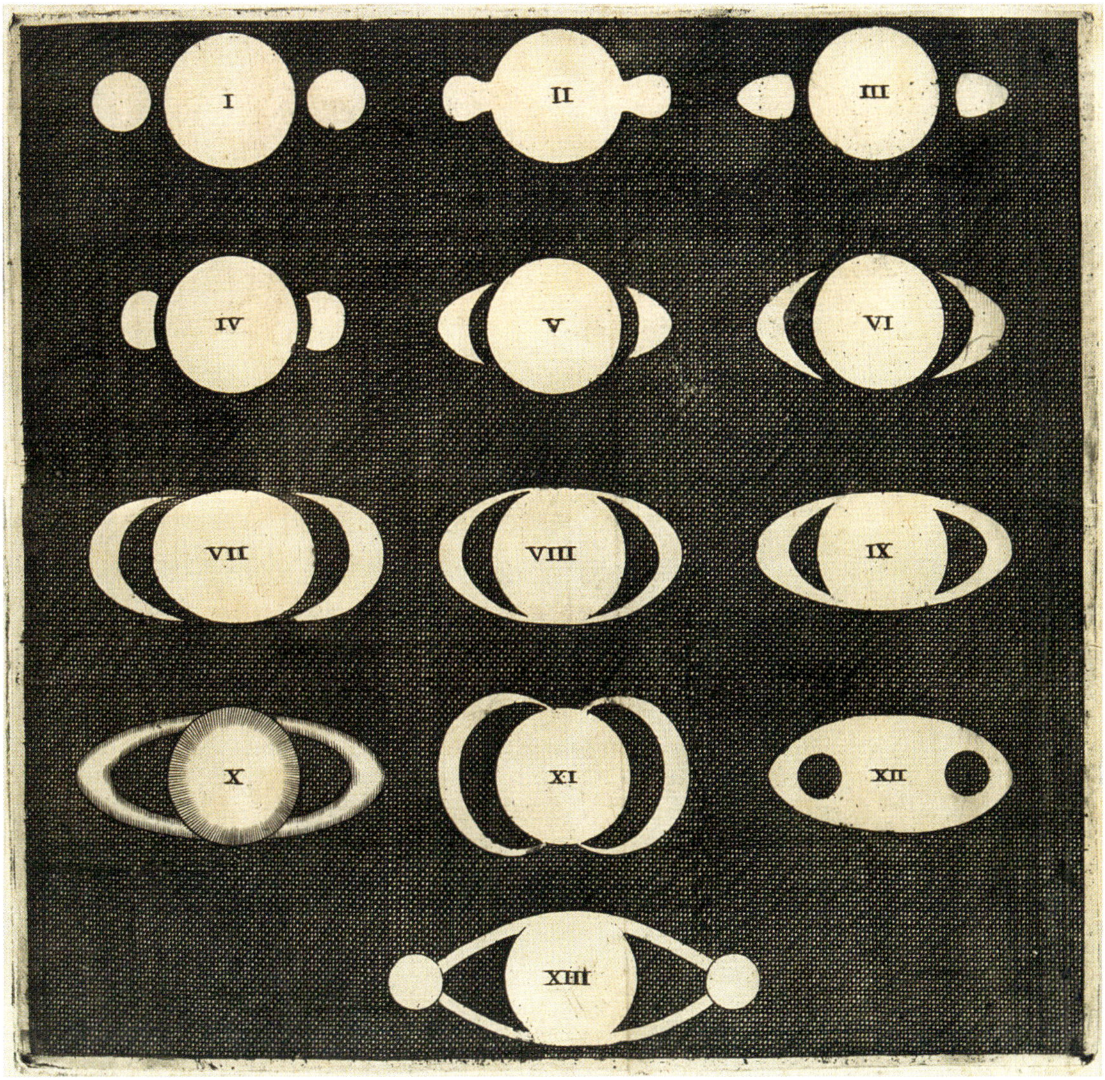

Fig. 4.2.5 - Different aspects of Saturn as perceived by early 17th century astronomers (C. Huygens, *Systems Saturnium*, 1659, p. 35).

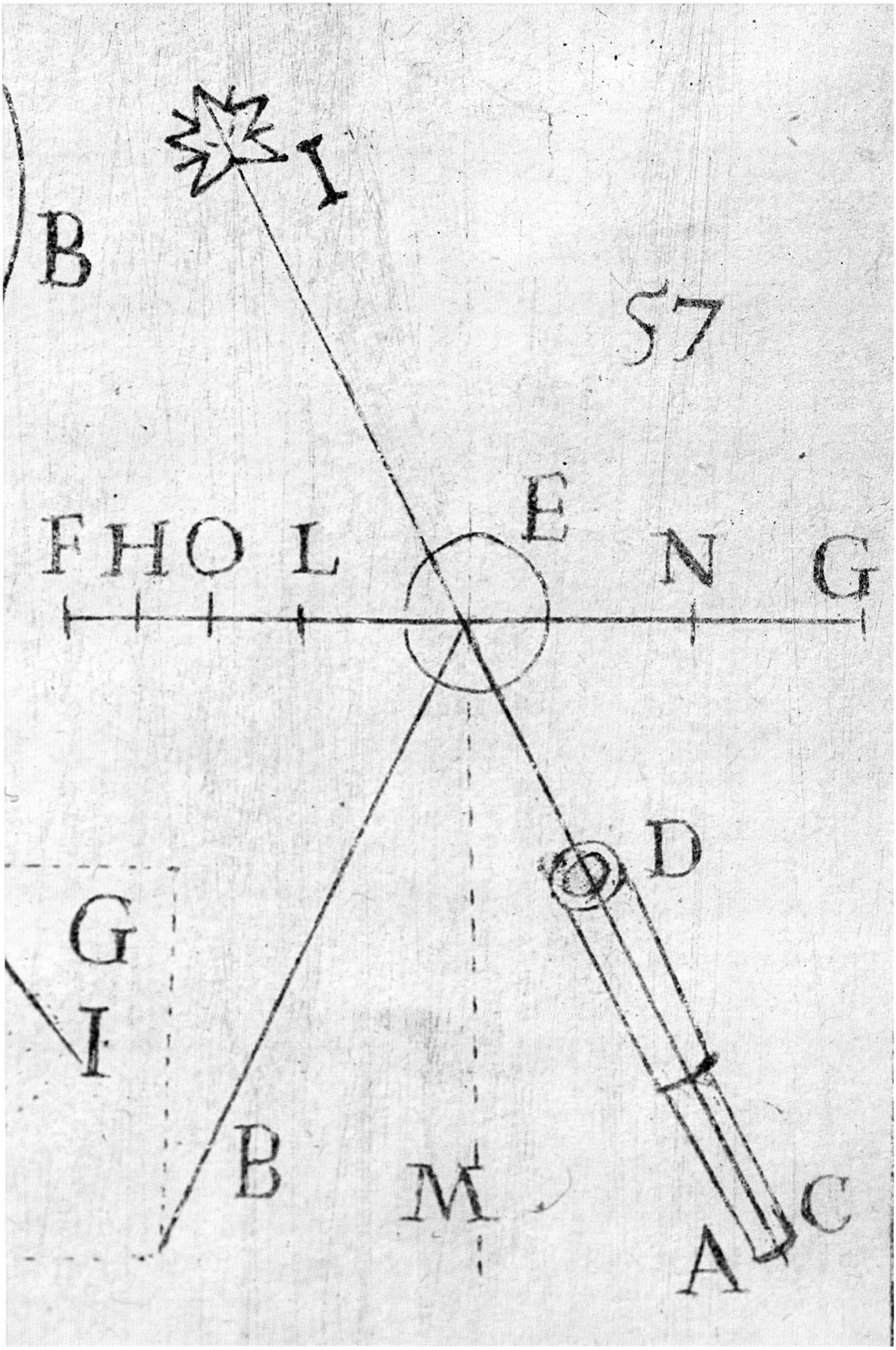

I
B
57
F H O L
E
N
G
D
G
I
B
M
C
A

this lack originated from an optical problem with Galileo's telescopes. The combination of a convex objective and a concave ocular produces a very small field of view. Indeed, Galileo's surviving telescopes, both with a magnification of about twenty times, have fields of view of about 15'. The field of view would have been even smaller for the telescope with a magnification power of about thirty times as is mentioned in *Sidereus Nuncius.* To complete a Moon drawing, Galileo had to assemble several observations of specific regions of the lunar disk. This process could have lead to the unintentional overlooking of the *Mare Serenitatis.*

Saturn: Degraded images of Saturn at a resolution of 2", 5" and 10" **(Fig. 4.2.4)** confirm that, in order to distinguish the planet's ring, a good quality telescope is essential. In particular, images at a resolution of 10" show a close resemblance to Galileo's descriptions and drawings of the planet. When the ring is less inclined to the observer, the parts of the ring close to the planet's disk appear very thin due to the effect of visual perspective. They tend to disappear, and the observer perceives two round bodies on each side of the planet's disk. When the ring is more inclined to the observer, the ring merges with the planet's disk, and the observer perceives two elliptic 'handles'. At the greatest inclination, the ring is partially covered by one of the polar regions of the planet. Since polar regions are slightly darker than the other parts of the planet's disk, the 'handles' may appear as protruding from one of the poles of Saturn. The human psychological predisposition to perceive symmetry explains why the protrusion of the 'handles' from both poles is also represented in Galileo's drawings.

The comparison confirms that Galileo observed Saturn with the same telescope he used to improve upon the Moon engravings for *Sidereus Nuncius.* Moreover, this comparison also brings to light that Galileo's ability as a telescope maker was not equalled by any other optician until the mid-17th century. Indeed, by examining the synoptic table of early observations of Saturn prepared by Christiaan Huygens (1629-1695) for *Systema Saturnium* (1659), it is apparent that several Italian and European astronomers had telescopes of a quality comparable to Galileo's **(Fig. 4.2.5)**. A considerable increase in quality to a resolution power of up to 5" or to 2" was only obtained by Campani and Huygens in the 1650's. (*G.S., M.Va.*)

Fig. 4.3.1 - Diagram of Galileo's micrometer (G.A. Borelli, *Teoricae Mediceorum Planetarum*, 1666).

Fig. 4.3.2 - Use of the helioscope (C. Scheiner, *Rosa Ursina*, 1630).

4.3 – Observing with Interactive Replicas of Galileo's Telescope

Some of the problems encountered by Galileo in his astronomical investigations, and the observation techniques he devised, can be more clearly understood by utilising interactive instruments in closed spaces, such as those of a museum or an exhibition. Each heavenly body, in fact, raised specific problems for Galileo, as regards recording his observations and, even more important, deducing from them objective data on which to formulate astronomical theories.

The micrometer: In the months following his discovery, Galileo systematically observed the satellites of Jupiter to determine their orbital radii and orbital periods, in order to compile astronomical almanacs predicting their positions with precision. He had in fact thought of a practical application for this discovery, using the satellites as a kind of celestial clock that could be employed for determining longitude in the open sea (an age-old problem solved only in the 18th century with the invention of mechanical chronometers that retained the necessary accuracy during navigation). To determine the observed positions of the satellites, Galileo invented a device – now called the "Galilean micrometer" – which, unfortunately, is known to us today only through a brief description by Giovanni Alfonso Borelli (1608-1679) **(Fig. 4.3.1)**. The micrometer consisted of a graduated measuring rod (or perhaps a grid) mounted on the side of the telescope and sliding along the tube. When the observer looks through the telescope with one eye while keeping the other fixed on the micrometer, the brain synthesises the two images, superimposing the micrometer on the telescope's field of view and thus allowing the distance of each satellite from the planet to be estimated. The micrometer was in fact graduated, with the apparent radius of Jupiter assumed as unit of measurement, and was calibrated on each occasion, by moving it further away or closer to the eye, on the disc of the planet observed through the telescope. Galileo's method can be clearly understood by aiming a replica of the telescope, fitted with a micrometer, at an image of Jupiter's system.

The helioscope: The first users of telescopes observed the Sun directly, without the aid of filters, with consequent damage, often irreversible, to their eyesight. Galileo instead adopted a method – devised by his pupil Benedetto Castelli (1587-1643) – consisting of projecting the image of the Sun through the telescope onto a sheet of paper placed at a distance of "four or five palms" (about one meter) from the eyepiece **(Fig. 4.3.2)**. On this sheet Galileo had previously drawn a circle of arbitrary size; he then adjusted the distance of the eyepiece so that the image projected by the Sun overlapped the circle exactly. At that point it was easy to record on the sheet, with great precision and no damage to the eyesight, the shape and position of the sunspots.

The satellites of Saturn: The reasons why Saturn appeared to Galileo in the form of a triple-bodied planet are clear today. The low resolution of the telescope, the reduced inclination under which Saturn's ring appeared to the observer on Earth in June 1610, as well as the two dark loops of empty space between the inner part of the ring and the planet, gave Galileo – and those who repeated the obser-

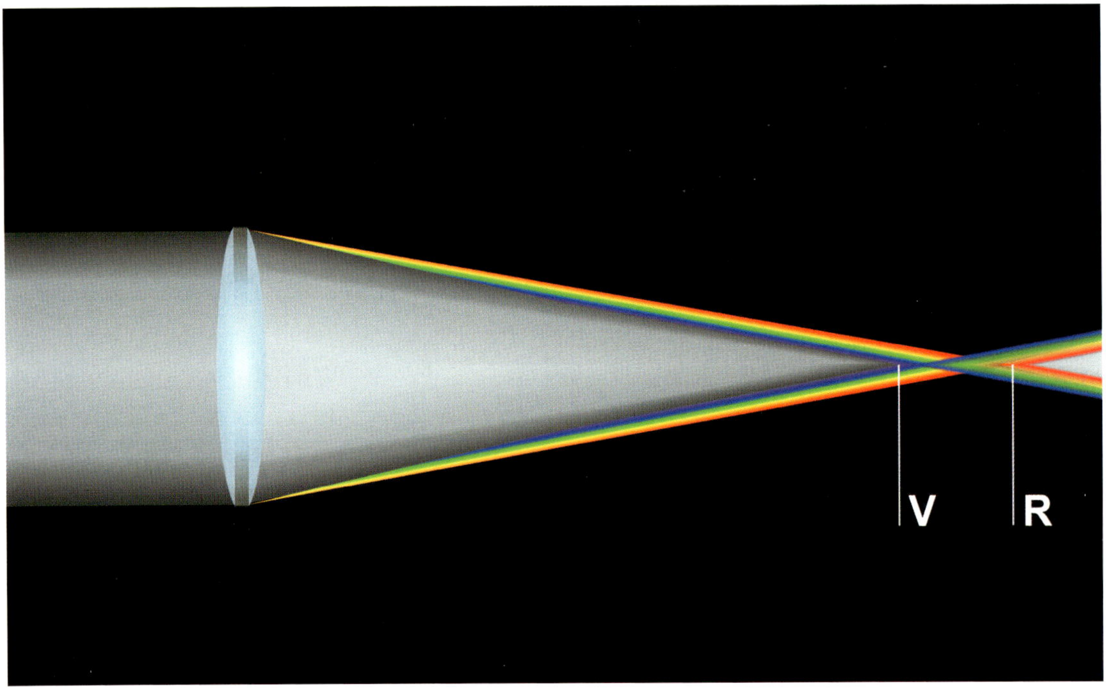

Fig. 4.3.3 - Behaviour of white light traversing a converging lens.

vations in the following months – the impression of seeing three distinct, separate heavenly bodies. When these factors are replicated – that is, an instrument of sufficiently low power is aimed at an image or three-dimensional model of Saturn with the ring only slightly inclined – the planet appears through the telescope to be made of three bodies.

Chromatic aberration and diaphragms: A beam of light that traverses a surface separating two different mediums – air and glass, for example – is refracted, that is, deviated. However, as demonstrated by Isaac Newton (1643-1727), ordinary light (the so-called white light) is made up of different colours and the angle at which it is refracted differs for each of them. This is why, in a converging lens, the focal point of violet light is closer to the lens than that of red light, while the other colours have focal points intermediate between these two **(Fig. 4.3.3)**. This phenomenon, known as "chromatic aberration", constitutes a severe problem in the telescopes – such as those of the 17th century – whose objective consists of a single lens. The impossibility of finding a single focal point for all of the colours provokes a phenomenon of iridescence that lowers the quality of the images. The problem was solved only in 1729, when the Englishman Chester Moor Hall (1704-1771) devised the first achromatic objective, composed of two lenses, one converging, the other diverging, each made of glass having a different index of refraction.

The effect of chromatic aberration becomes particularly evident when a telescope is aimed at a backlit image. It is however interesting to note that, even without recurring to additional lenses, the aberration effect can be reduced by placing in front of the objective a diaphragm, consisting of an opaque disc with a small hole in the centre. This was in fact the solution introduced in Galileo's telescopes and those of his successors.

Magnification test: In the *Sidereus Nuncius* Galileo specifies that, to observe what he has discovered, one must have a telescope of good quality with magnification of at least 20. He then explains to the reader a method by which "anyone can, with little effort, make certain of the instrument's magnification power". The method consists of preparing two circles (or squares) of paper whose diameters (or sides, in the case of squares) have a ratio of 1 to 20. When the two circles are observed contemporaneously – the smaller one through the telescope and the larger one with the free eye – the images will overlap, exactly as in the case of the micrometer. If the telescope in question really has a magnification of 20, the two circles will have the same apparent size and, superimposed one upon another, will appear to be a single circle. *(P.D.S.)*

IOANNES KEPLERVS

Chapter 5:

From Galileo to Newton

Paolo Del Santo, Giorgio Strano

5.0 - Introduction

For several years, Galileo maintained an outstanding leadership in telescope production. It is a matter of fact that, after the printing of the *Sidereus Nuncius*, Kepler **(Fig. 5.0.1)**, Court Mathematician for the Holy Roman Emperor Rudolph II, complimented Galileo on his remarkable discoveries. However, in his letter, published in Prague and in Florence with the title *Dissertatio cum Nuncio Sidereo* (1610), he also expressed a few problems he found with Galileo's discoveries that he wanted to resolve by telescopic observation. For several months, Kepler searched for a good telescope to use to observe, in particular, the four new planets orbiting around Jupiter. When he finally found a suitable instrument on loan, it was the one that Galileo had made for the Elector of Cologne. Only by using this telescope Kepler was able to observe the four satellites of Jupiter for a few days during August 1611. Immediately thereafter, he confirmed this celestial discovery in the booklet *Narratio de observatis a se quatuor Iovis satellitibus erronibus.*

As another example, after reading the *Sidereus Nuncius*, Nicolas Claude Fabri de' Peiresc (1580-1637) had the idea to precisely determine the orbital periods of the satellites of Jupiter. He was absolutely convinced that he could beat Galileo at his own game. Therefore, he ordered an artist, Jean Chalette, to prepare the frontispiece for his book on the subject **(Fig. 5.0.2)**. In his attempt to accomplish this astronomical work, Peiresc used a telescope made by Jacob Metius, one of the three alleged fathers of the instrument. However, after Galileo's condemnation by the Holy Roman Inquisition in 1633, Peiresc wrote to Galileo to offer his solidarity. In addition, he pointed out that Metius' instrument that he possessed in 1610-1611 allowed him to see the four satellites of Jupiter, but did not allow him to conduct any serious scientific work. For this reason, in 1634, Peiresc asked Galileo if he was still able to burden the task of producing and sending him a good telescope.

Fig. 5.0.1 - Portrait of J. Kepler (Florence, Uffizi Gallery, Gioviana Collection, n. 248).

A Aix 1611

These two examples form additional evidence of the uncontested quality of Galileo's instruments for over more than twenty years after his exceptional astronomical discoveries of 1609-1611. However, it may appear puzzling that Galileo was able to improve upon early spyglasses from the Netherlands in a couple of years, and nobody else was able to improve upon Galileo's telescope for more than twenty years. Even more astonishing is the fact that, after the quick development of the telescope between 1609 and 1610, Galileo himself was not able to further improve upon the instrument. Things seem to suggest that, in 1610, he had at hand an optical device impossible to be further developed, which was, at the time, intrinsically true. (*G.S.*)

5.1 - Conceiving New Telescopes

The early evolution of the telescope was the product of a paradox. On the one hand, in 1611, the telescope that Galileo improved upon by an empirical process gained an universal success. On the other hand, in 1611, a new type of telescope emerged from the theoretical approach to the law of refraction as applied to combinations of lenses. Yet, it was of a design impossible to be made with the lens quality of the day. This paradox appears intriguing by considering that the new type of telescope had, in theory, valuable astronomical advantages over Galileo's.

Noticeably, the combination of a convex objective and a concave ocular in the Galilean telescope had advantages; the first of which was producing erect images. Such a characteristic made this type of telescope useful for both terrestrial and astronomical use. In addition, the telescope's image formation was not highly sensitive to incorrect focusing. Slight shifts of the ocular out of the ideal position produced zooming effects that did not overly affect the clarity of the perceived image. Also slight tilting of the objective or the ocular, or a slight dislocation of the latter away from the telescope's axis produced minimal distortions.

The relevant limitation of the Galilean telescope consisted in the concave ocular's effect to reduce the instrument's field of view, which was a direct outcome of the magnification. Indeed, there were two ways to increase telescopic magnification. First, the focal distance of the objective could be increased, meaning that the objective must be placed farther away from the ocular. Second, the focal distance of the ocular could be reduced, meaning that the ocular must have a more pronounced concavity. As a result, by looking through a high power Galilean telescope, the observer saw the objective's diameter reduced both for its increased distance from the

Fig. 5.0.2 - J. Chalette, Draft for the frontispiece of Peiresc's planned book on the satellites of Jupiter (Carpentras, Bibliothèque Inguimbertine, Ms 803, f. 285).

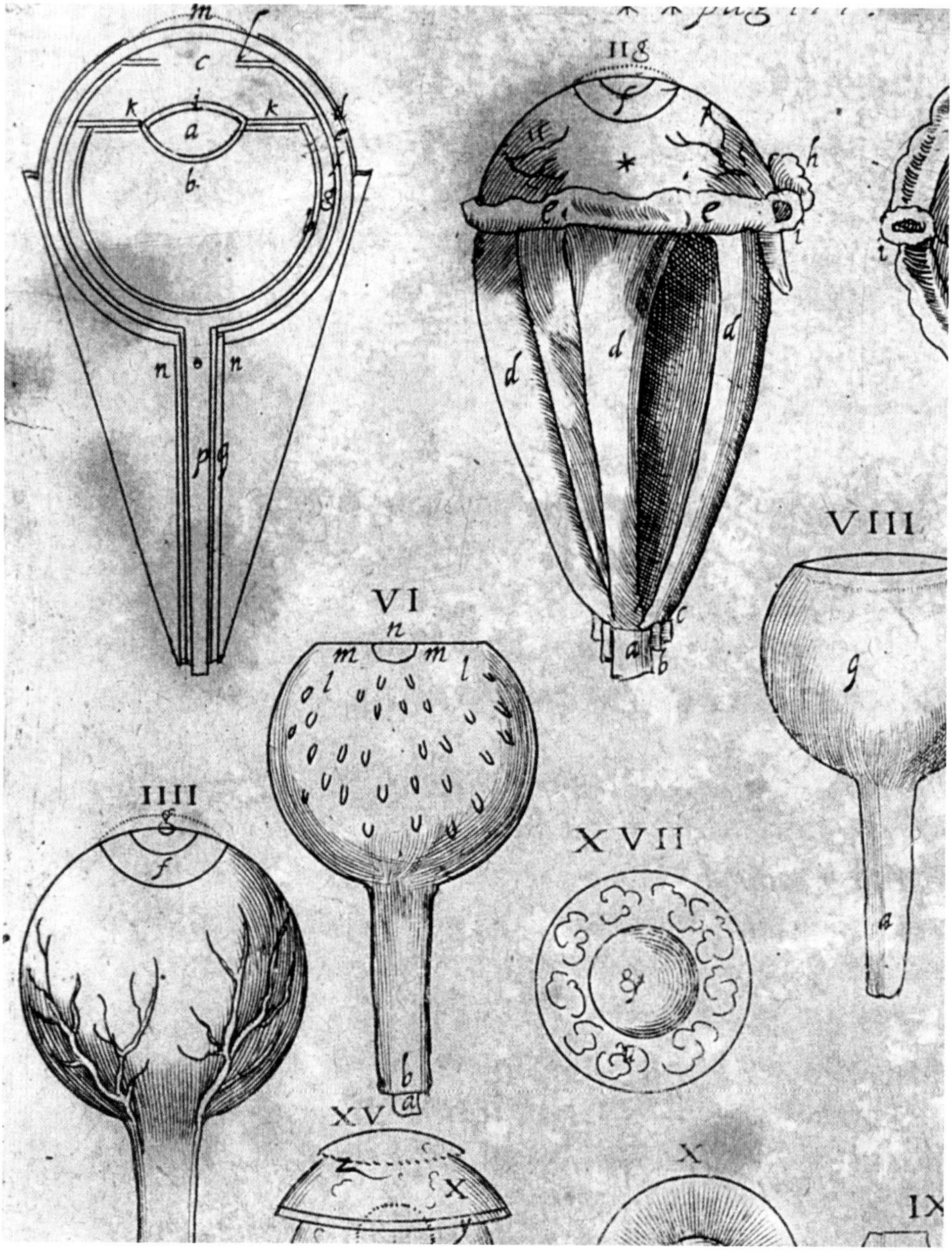

Fig. 5.1.1 - The human eye as an optical device (J. Kepler, *Ad Vitellionem Paralipomena*, 1604).

eye and for the increased reducing effect of the ocular. This phenomenon explains why, after such rapid progress, Galileo was unable to further improve upon the instrument. The magnification of thirty times he mentioned in the *Sidereus Nuncius* was the intrinsic limit for his telescope.

The solution to overcome this limit was theoretically not far off. In 1604, Kepler revealed his great interest in optics by publishing *Ad Vitellionem Paralipomena*. The book began with the geometrical demonstration of the theory of the rectilinear propagation of light, and the law of reflection. The demonstration was essential to determine the apparent diameters of the Sun and the Moon, and to study eclipses. However, Kepler also examined refraction of light rays traversing the surface between two transparent media. In particular, he introduced refraction between crystalline ether and air, in order to deal with the so-called 'astronomical refraction' (partially explained by Ptolemy and Tycho). Kepler also added a brief study of the human eye, considered as an optical device into which refraction played a key role in image formation **(Fig. 5.1.1)**. This study involved the behavior of light rays passing through transparent spheres, and convex and concave lenses.

Kepler continued to study optics, and in 1611, he published a new book on the subject in Augsburg. This book, *Dioptrice*, represented a thorough geometrical study of light rays refracting through glass cubes and prisms, through plane, concave and convex surfaces, and through lenses of several forms. Kepler also studied the relation between the position of a body and the position of its image produced by the lens **(Fig. 5.1.2)**. Finally, he studied the possibility of using the image produced by a first

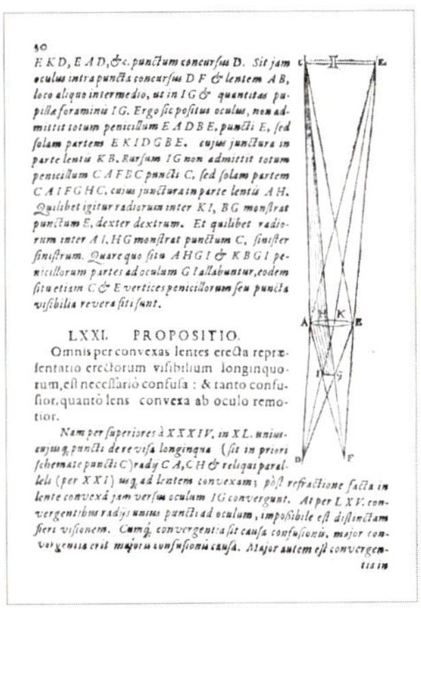

30

E K D, E A D, &c. punctum concursus D. Sit jam oculus intra puncta concursus D F & lentem A B, loco aliquo intermedio, ut in I G & quantitas pupillae foraminis I G. Ergo sic positus oculus, non admittit totum penicillum E A D B E, puncti E, sed solam partem E K I D G B E. cujus junctura in parte lentis K B. Rursum I G non admittit totum penicillum C A F B C puncti C, sed solam partem C A I F G H C, cujus junctura in parte lentis A H. Quilibet igitur radiorum inter K I, B G monstrat punctum E, dexter dextrum. Et quilibet radiorum inter A I, H G monstrat punctum C, sinister sinistrum. Quare quo situ A H G I & K B G I penicillorum partes ad oculum G I allabuntur, eodem situ etiam C & E vertices penicillorum seu puncta visibilia revera siti sunt.

LXXI. PROPOSITIO.

Omnis per convexas lentes erecta repraesentatio erectorum visibilium longinquorum, est necessariò confusa : & tanto confusior, quantò lens convexa ab oculo remotior.

Nam per superiores à XXXIV. in XL. uniuscujusq; puncti de re visa longinqua (sit in priori schemate puncti C) radij C A, C H & reliqui paralleli (per XXI) usq; ad lentem convexam; post refractione facta in lente convexâ jam versus oculum I G convergunt. At per LXV. convergentibus radijs unius puncti ad oculum, impossibile est distinctam fieri visionem. Cumq; convergentia sit causa confusionis, major convergentia erit majoris confusionis causa. Major autem est convergentia in

Fig. 5.1.2 - Determining the image formation of a bi-convex lens (J. Kepler, *Dioptrice*, 1611, p. 30).

44

Sit enim A B visibile, C D, E F lens ab oculo remotior. Sit K punctum concursus. Si ergo imago ipsius A B evertitur hac unâ lente: punctum ubi imago apparet eversa, erit ultra K remotius à lente per L X X V. Sit ille locus L. & quia species ipsius lentis E F, cumq; eâ imago eversa ipsius A B, debet everti denuò per aliam lentem, quae sit G H, imago verò rei A B eversa, comprehenditur lineis A D F L, B C E L: necesse est igitur lentem G H, esse ultra L. per LXXVI. Fuit verò L à lente E F remota ultra K punctum concursus. Ergò G H lens secunda multò longius removebitur, ultra K ejus punctum concursus: ut F L G, E L H venientes ab extremitatibus rei, secundam refractionem in G, H. passi tandem iterum coëant, & cogantur ad oculum in I.

Deniq; haec imago minor est re visibili. Nam primùm species ipsius E F (eorumq; quae per eam videntur) eversa per lentem G H, & distincta apparens, erit minor in I, per XXCV. Sed per eandem, oculo in L constituto, ipsum etiam visibile A B per lentem C D eversum, minus occupare spacium videtur in lente, quàm pro suâ magnitudine. Quia L non potest esse proximum ipsi K puncto concursus, ne nimia sit confusio. L enim proximum esse debet puncto distinctionis, ut & I. Gemino igitur nomine visibile A B repraesentatur parvum.

XXCIIX. PROPOSITIO. PROBLEMA.

Duobus convexis pingere visibilia super papyro situ erecto.

Problema diu quaesitum. Habeant igitur convexa ut prop. XXCVII. scilicet ut lens propior papyro sit ultra K

Fig. 5.1.3 - Scheme of the Keplerian telescope (J. Kepler, *Dioptrice*, 1611, p. 44).

lens as the object for the image produced by another lens. If both the objective and the ocular lenses had a convex shape, the resulting optical device — nowadays known as the 'Keplerian' telescope — was capable of magnifying at a distance **(Fig. 5.1.3)**.

Introducing the convex ocular had the remarkable advantage of overcoming the small field of view of the Galilean telescope. Theoretically, this allowed the new telescope to reach a magnification power far beyond the limit of thirty times. However, Kepler did not succeed in making this new telescope, and he had to use a Galilean instrument for his astronomical observations. Unfortunately, by combining two convex lenses the optical aberrations in image formation considerably increased, and the new instrument was extremely sensitive to focusing and to errors in the ocular's alignment. Moreover, the ocular amplified distortions and imperfections of the objective, and produced colour fringes around the object being observed. These reasons may explain why it was the Galilean, and not the Keplerian type of telescope that appeared in the Netherlands in 1608: any telescopic combination of convex spectacle lenses would have produced poor quality images. Another reason may be the fact that the new type of telescope created inverted images of the objects being observed,

Fig. 5.1.4 - E. Torricelli, Objective lens, 1647, reused in a 18th century metal telescope (IMSS, inv. 2554).

making the device completely useless for military purposes. To overcome such a fault — irrelevant from the astronomical point of view — Kepler knew that it was indispensable to introduce a third convex lens between the objective and the ocular to right the image. Such a third lens would have increased optical aberrations already present in the two-lens system.

In fact, the early evolution of the telescope proceeded along the path set out by Galileo. The main concerns of his followers were the refinement of the techniques for producing optical glass and for grinding and polishing lenses. Ippolito Francini (1593-1653) took over Galileo's trade, and continued to make telescopes after the death of his master. Francesco Fontana (*c.* 1585-1656), made telescopes in Naples, that he also used for his astronomical observations. Finally, Evangelista Torricelli (1608-1647) recuperated the Florentine leadership in making lenses. Unfortunately, optical components that still survive in museums are not completely useful to understand to what degree the Keplerian telescope was adopted by Galileo's followers. Indeed, the usual scientific tendency to reuse good optical components (the so-called "cannibalization" process of scientific instruments) created a sort of artificial selection of antique lenses. Usually, only the most important and valuable part of an old telescope, the objective lens, was conserved and reused in order to create a new telescope **(Fig. 5.1.4)**. Original ocular lenses, small and easy to be made, were often replaced and thrown away.

For certain, the complete release of the Keplerian telescope required the mastering of a new level of precision in lens-making techniques. This evolution led to the creation of a new professional figure, the specialised optician. Once this social figure was affirmed the fate of the Galilean telescope was sealed. After opening the way for a complete change in the understanding of the Universe, the telescope trade re-transformed the Galilean instrument into a simple eyeglass. With a reduced magnification of two or three times, it became useful in assisting the wealthy and the nobility in watching theatrical representations. (*G.S.*)

5.2 - Telescope-Making in the Mid-17th Century

Notwithstanding the downfall of the Galilean instrument, Italy remained at the centre of telescope making for half a century. In particular, it was in Rome that Eustachio Divini (1610-1685) from San Severino Marche and Giuseppe Campani (1635-1715) from a small village near Spoleto established their workshops. Neither Divini nor Campani were natural philosophers and they did not need to prepare instruments for conducting personal scientific research. Both began their careers as clock makers, and evolved into telescope and microscope makers. Due to their ability, they started the professional figure of the specialised optician, someone able to prepare powerful instruments on the request of natural philosophers and public institutions. This evolution had a strict connection to the development of the Keplerian telescope and its 'terrestrial' variants, and to the refining of well-developed grinding and polishing techniques.

Little information is available regarding how Galileo procured the lenses for his telescopes. Present historiography suggests that perhaps he selected them from the lens-market, or he educated a few glass makers to fulfil his needs, or he made the lenses himself. Whichever the case, at the very beginning of the development of the telescope, Galileo tried to make his own lenses with basic methods. On the blank part of a letter from one of his correspondents, dated 23 November 1609, he wrote a shopping list including "No. 2 cannon balls. Organ pipe made of tin. German flat glasses. [...] Tripoli. [...] Colophony." **(Fig. 5.2.1)**. Such items testify to Galileo's attempt to gather the materials necessary to prepare the tube and the lenses for his most advanced telescope. In particular, he was planning to use cannon balls as moulds to shape concave oculars, and iron or rock bowls as moulds for convex objectives. He was also planning to use adhesive, abrasive, and polishing substances to grind the lenses.

It is uncertain whether Galileo created special tools or machines for grinding and polishing lenses. However, such tools and machines became an essential part of the workshops of the specialised optician. As Carlo Antonio Manzini (1600-1677) refers to in his book on practical optics, *L'occhiale all'occhio, dioptrica pratica* (1660), the construction of lenses intended for a mid-17th century telescope went through several phases.

Firstly, the optician had to search for very good pieces of glass useful to make the large objective lens. Usually, opticians bought several plates of glass made by glass makers, especially by those in Venice. All the plates where examined carefully to detect possible imperfections (bubbles, inclusions, faint stripes, chips and micro-cracks) using a magnifying glass or a microscope. Due to this process of analysis, the majority of the plates were rendered useless and discarded. As Divini wrote to Manzini, in a series of 24 plates he received from Venice, only three were found suitable for producing lenses.

Secondly, the glass plate had to be given the general form of the lens. Iron dividers were used to mark out a circumference on the plate. Thereafter, the excess

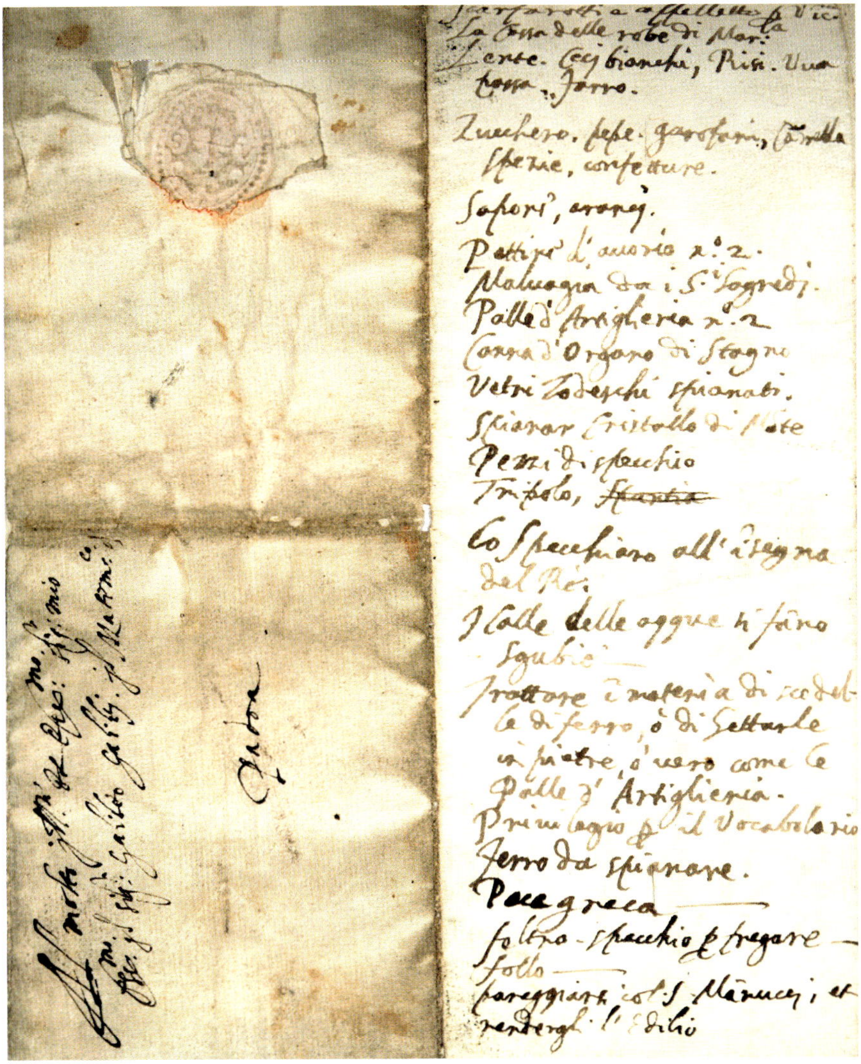

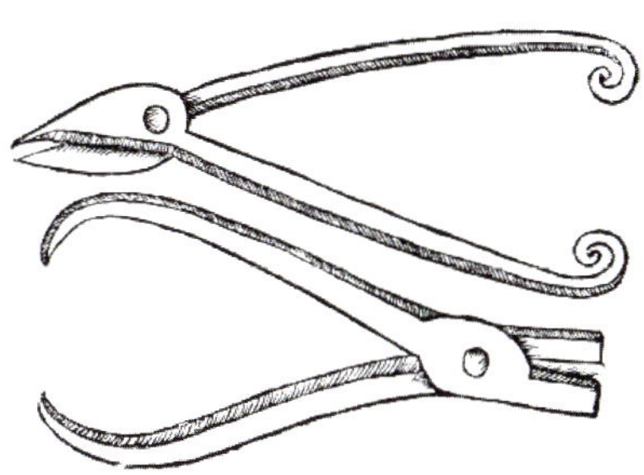

Fig. 5.2.1 - Galileo's shopping list, *c.* 1609 (BNCF, Ms. Gal. 88, ff. 106r-107v).

Fig. 5.2.2 - Cutters and pinchers to work glass plates (C. A. Manzini, *L'occhiale all'occhio*, 1660, p. 120).

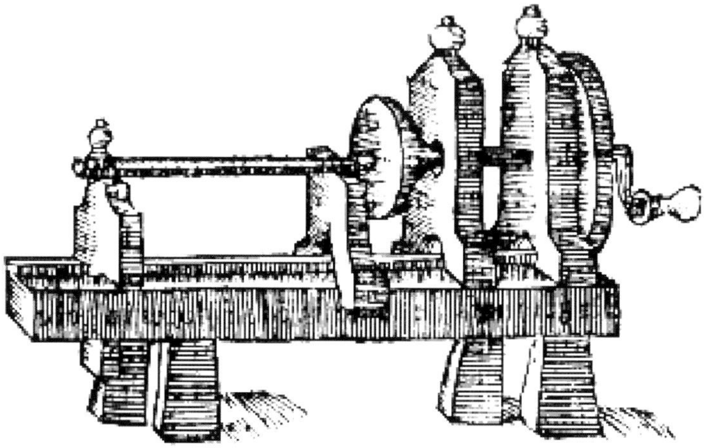

glass was removed by a diamond point and other instruments such as cutters and pincers **(Fig. 5.2.2)**. To avoid any excessive stress on the glass, opticians did not file a lens' edge after cutting. This explains why surviving lenses from this date have heavily chipped edges. Usually, the rough edge of a lens was concealed by using cardboard rings as diaphragms.

Thirdly, the lens was ground on a mould. In order to obtain a considerable magnifying power, the objective lens of a telescope would have to have a very long focal length. As a consequence, both spherical surfaces of the lens would have long curvature radii. Cannon balls and iron or rock bowls would have been of no aid for this specialised production. Therefore, opticians made special metal moulds with a lathe **(Fig. 5.2.3)**. To grind the lens, opticians fixed it on a proper handle by a strong adhesive. This adhesive was colophony; it had the advantage of being easily removed by heating at the end of grinding. The lens was then rubbed on the mould with the interposition of abrasive powders. During the process the abrasive powder used was progressively finer, in order to obtain a very smooth (still opaque) surface. The substances which opticians used were coarse and fine silicon sand and emery. The grinding process was often 'mechanised' by using several types of lathes **(Fig. 5.2.4)**.

Lastly, to complete the preparation, both surfaces of the lens had to be as transparent as possible, that is, without even a minute scratch produced by the erosive action of the abrasives. Therefore the surfaces of the lens were polished by employing soft fabric with the addition of a very fine abrasive powder, the so-called 'tripoli', and tin oxide.

The considerable advancements in lens making techniques created the possibility to design telescopes with additional lenses. This lead to overcoming the inversion of the telescopic image produced by the basic Keplerian telescope. In particular, Campani invented a special lens-doublet to correct such an alleged fault, and created a four-lens 'terrestrial' telescope. The doublet, called the 'erector' on the bases of its optical function, was made of two equal convex lenses placed between the objective and the ocular of the telescope. Campani's rival, Divini, studied other optical combinations, involving both convex and concave lenses in considerable numbers. His production was not as 'standardised' as Campani's, and one of his telescopes included up to seven different lenses.

The specialised procedures and apparatus were quite complex, costly, and required trained and skilled craftsmen. Italian opticians became the leaders in telescope production. Campani in particular sold his telescopes all over Italy. In addition, he became the privileged furnisher of telescopes for the observatory of Paris, built be-

Fig. 5.2.3 - Lathe for preparing moulds (C. A. Manzini, *L'occhiale all'occhio*, 1660, p. 223).

Fig. 5.2.4 - Lathe for grinding and polishing lenses (C. A. Manzini, *L'occhiale all'occhio*, 1660, p. 158).

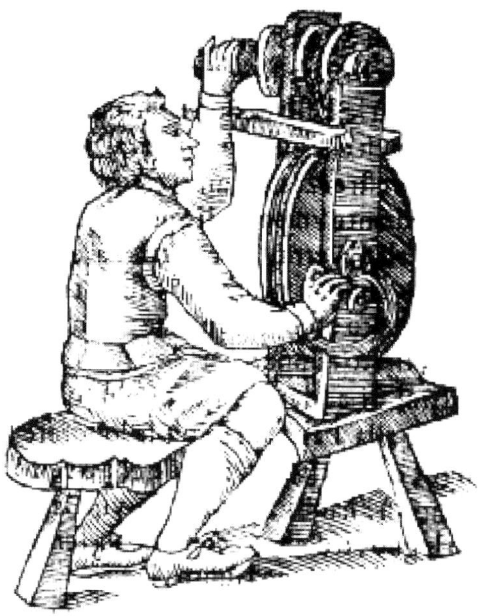

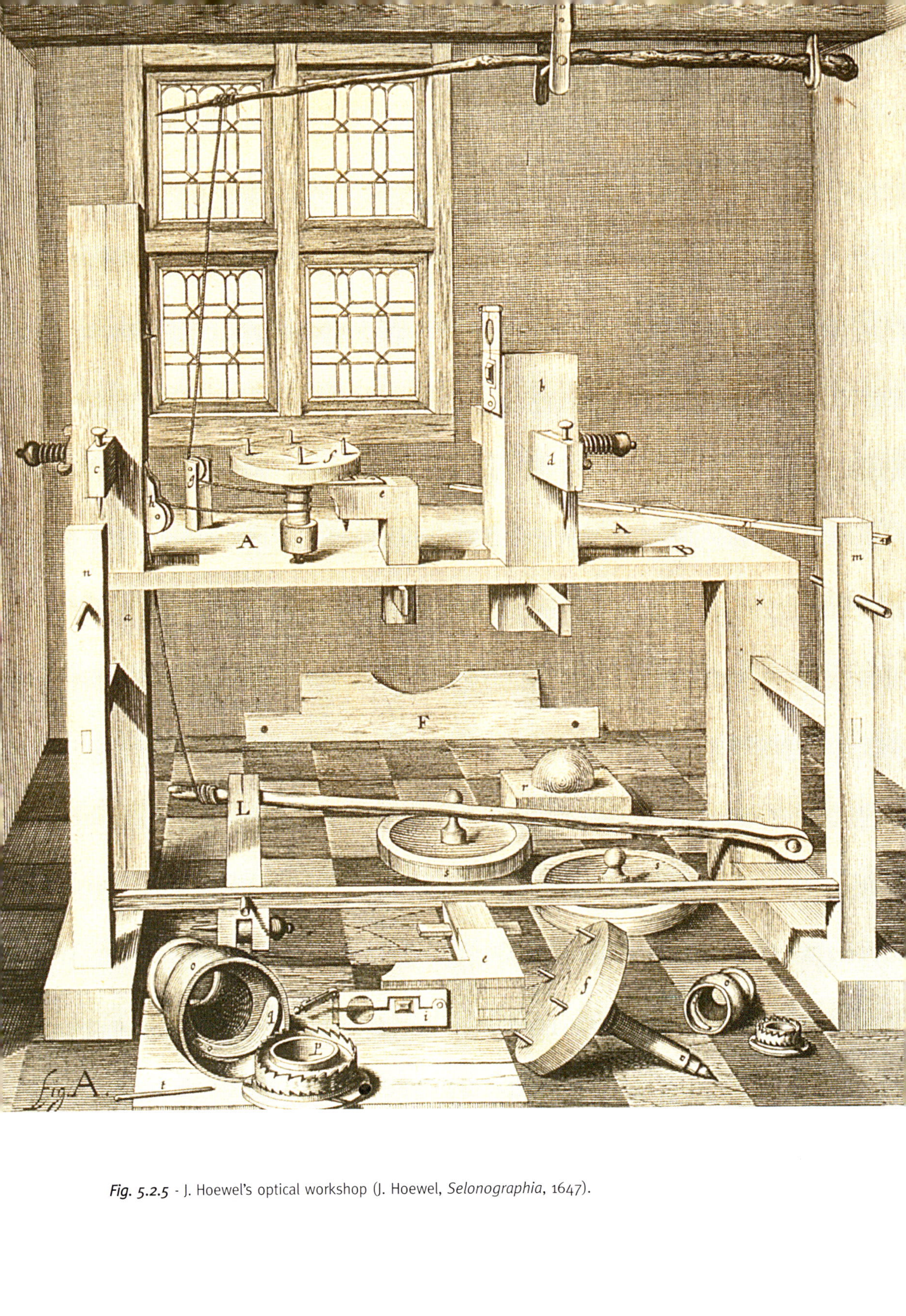

Fig. 5.2.5 - J. Hoewel's optical workshop (J. Hoewel, *Selonographia*, 1647).

tween 1667 and 1671, and directed by Giovanni Domenico Cassini (1625-1712). Thanks to Campani's telescopes, Cassini was able to detect the shadows of the Galilean satellites of Jupiter on the surface of the planet.

Also a few European natural philosophers and astronomers began to emulate the Italian products. For instance, at a scientific level, Huygens designed a new type of ocular that he used to discover the first satellite of Saturn and to identify the ring encircling the planet. At a more advanced technical level, the considerable funds that Johann Hoewel (1611-1687) received from his breweries, allowed him to organise a complete optical workshop to produce extremely long telescopes **(Fig. 5.2.5)**. (*G.S.*)

5.3 – The Evolution of the Telescope

Already the first constructors of telescopes, starting with Galileo himself, had empirically realized that the effects of chromatic aberration in a lens substantially diminished when the ratio between the focal length and the diameter of the objective was increased. This helps to explain why the objectives in 17th century telescopes not only had relatively long focal lengths but were also, as a rule, fitted with a diaphragm to reduce their aperture.

The focal length required to limit the effects of chromatic aberration is not proportional to the diameter of the objective, but to its square. If, for example, an objective with an aperture of 2 centimetres shows no appreciable chromatic aberration starting from focal lengths of around 75 centimetres, an objective having triple the diameter, i.e., 6 centimetres, must have a focal length at least nine times greater (= 3^2), that is nearly 7 meters! This circumstance profoundly conditioned the subsequent evolution of the telescope. Progressive increase in the diameter of the objectives led to an extraordinary increment in the length of telescopes, the practical limits of fabrication being reached within a few decades. While toward the middle of the 17th century the dimensions of a large telescope were around 4.5 meters in length with about 5 centimetres of aperture, some twenty years later astronomers were using instruments with apertures of 10 to 15 centimetres and focal lengths ranging from 10 to 45 meters **(Fig. 5.3.1)**.

It was thus no coincidence that, in the 1660s and 1670s, opticians and mathematicians devoted themselves to searching for alternative solutions based on the use of mirrors, which, operating by reflection, are not affected by chromatic aberration. In 1663, the Scottish astronomer James Gregory (1638-1675) devised an instrument consisting of a primary concave parabolic mirror, serving the function of objective, and a secondary elliptical mirror, also concave. The secondary mirror was positioned in front of the focus of the primary one and reflected the optical beam backward, towards an eyepiece placed behind the primary mirror, which had a hole drilled in the centre. The Gregorian reflecting telescope, which furnished erect images, met with great success in the 18th century **(Fig. 5.3.2)**, thanks especially to the famous Scottish optician James Short (1710-1768).

In 1668, Isaac Newton (1642-1727) also designed a telescope that utilised mirrors. To a concave parabolic objective he combined a secondary plane mirror, inclined by 45°, which served to reflect the optical beam laterally, outside of the telescope's tube, where the eyepiece was positioned. This first instrument, now lost, was followed by a second one, still existing today, which was presented to the Royal Society of London in 1672. In that same year the Frenchman Laurent Cassegrain (*c.* 1629-1693) proposed still another model of telescope with mirrors, again with a concave parabolic objective, in which the secondary mirror, a convex hyperbolic one, was placed in front of the focus of the primary. As in Gregory's telescope, this model too reflected the optical beam to an eyepiece placed in back of the primary mirror.

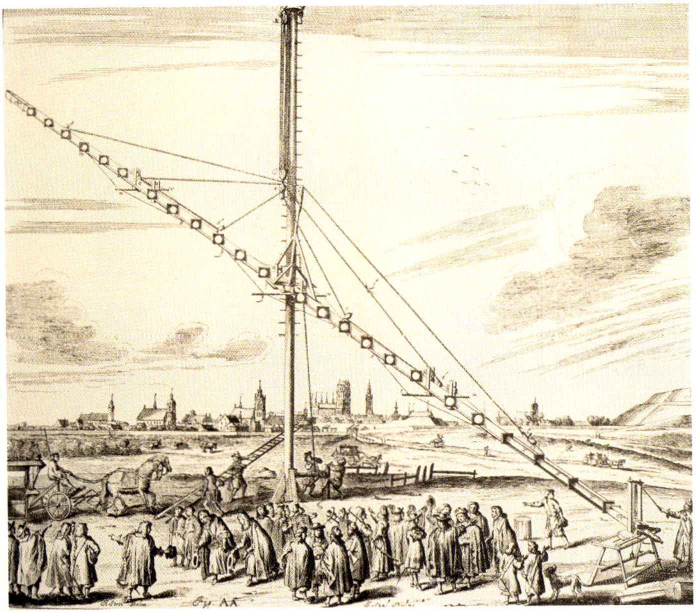

Fig. 5.3.1 - Telescope 150 feet (45 m) long built by J. Hoewel at Danzig (J. Hoewel, *Machina Coelestis*, 1673).

Fig. 5.3.2 – Optical production of L. Selva, including various Gregorian telescopes (L. Selva, *Sei dialoghi ottici teorico-pratici*, 1787).

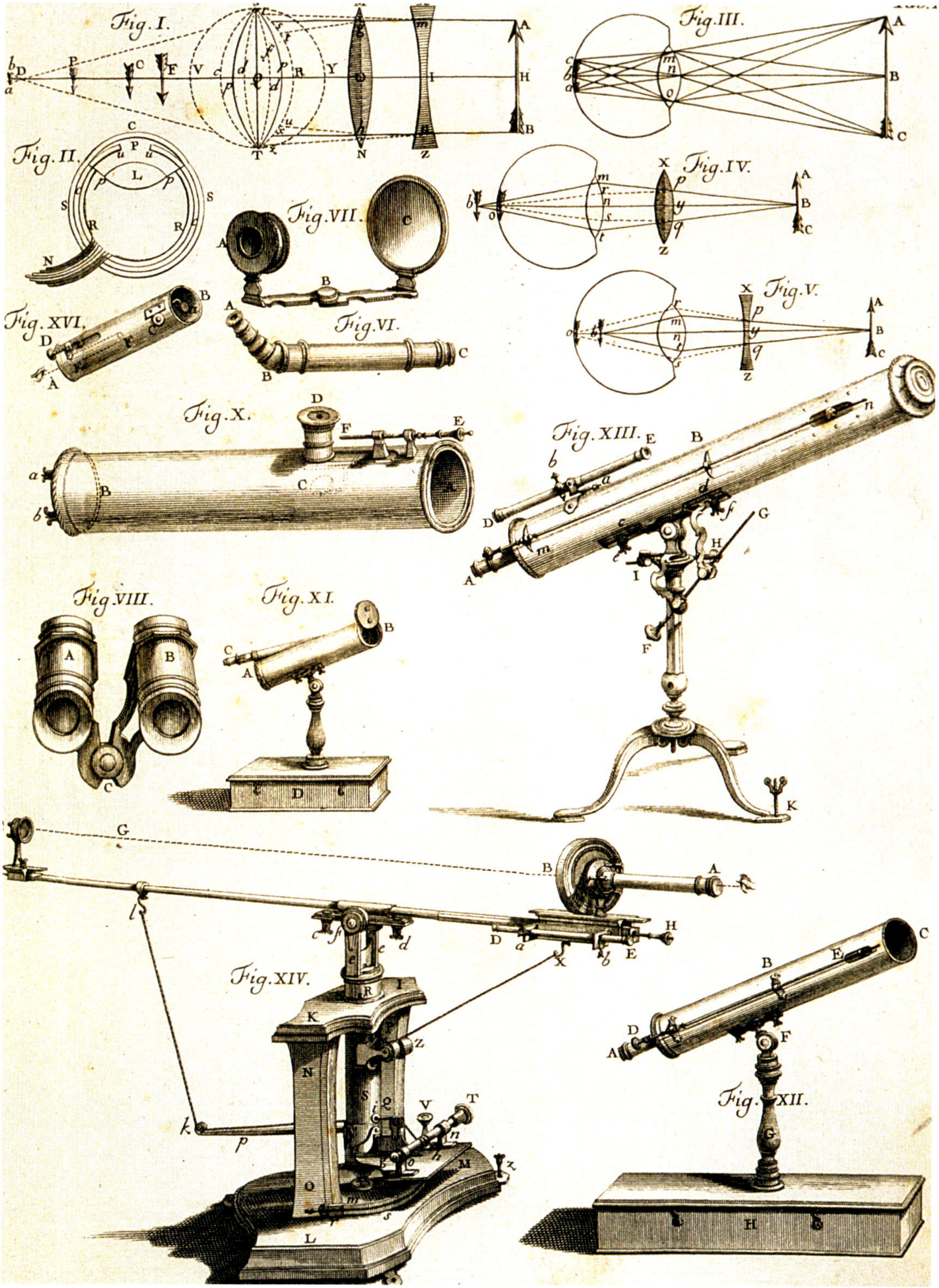
Fig. I.
Fig. II.
Fig. III.
Fig. IV.
Fig. V.
Fig. VI.
Fig. VII.
Fig. VIII.
Fig. X.
Fig. XI.
Fig. XII.
Fig. XIII.
Fig. XIV.
Fig. XVI.

The first truly efficient reflecting telescope was a Newtonian one with about 6 inches (approximately 15 cm) of aperture, presented in 1722 to the Royal Society by the Englishman John Hadley (1682-1744). With a length of only 1.6 meters, its performance virtually equalled that of Huygens' refracting telescope — having one lens as objective — which was 123 feet (over 37 m) long. Although many of these, some of large size, were built during the 17th century, what prevented the success of the reflecting telescopes (which now constitute 100% of large professional telescopes) was the poor reflecting power of the mirrors of the time, made of a copper and tin alloy called "*speculum*", which reflected only about 60% of the incident light. This problem was solved in 1856, when the process of silver-plating was introduced, allowing the utilisation of mirrors made of glass covered with a very fine layer of the purest silver.

The tube too, which supports the optical components and keeps them aligned, underwent significant transformation over the course of time, becoming standardised at the conclusion of the pioneering stage. With the increase in size, the use of metal (lead, tin, etc.) for the tubes, as in the earliest telescopes, was soon abandoned. In addition, the tubes became telescopic, built of several sections sliding into one another, so as to reduce the size of the instrument when not in use. The preferred material became pasteboard (and more rarely wood), light in weight but able to ensure the necessary rigidity. The secondary sections were often covered in marbled

Fig. 5.3.3 – Diagram of Newtonian telescope in which a prism is used in place of the secondary plane mirror (I. Newton, *Optice*, 1740, t. V, fig. 29).

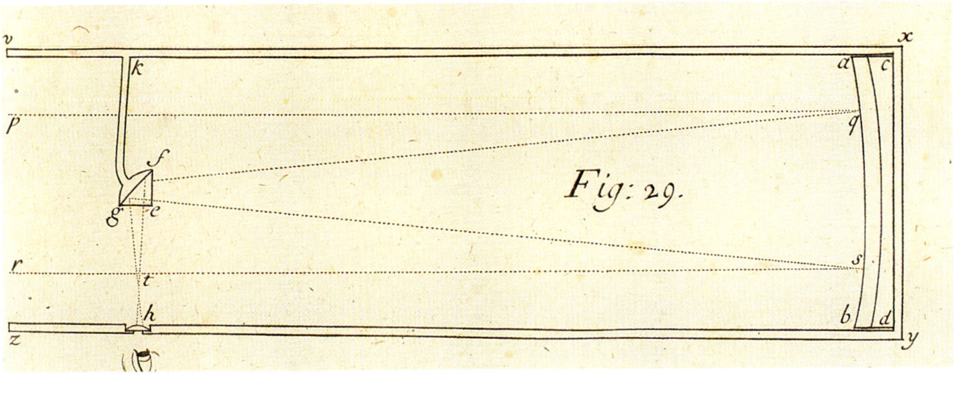

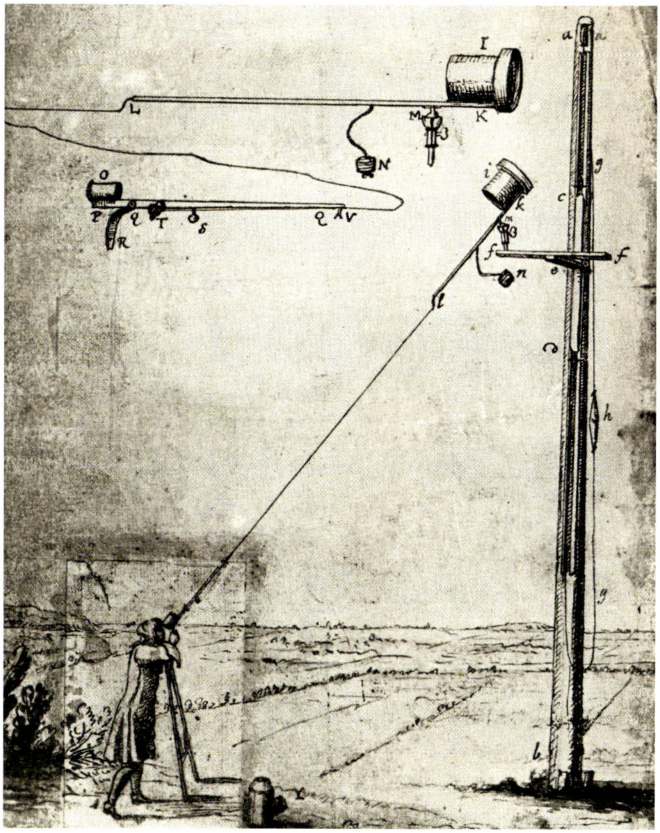

Fig. 5.3.4 – "Aerial" telescope 123 feet long by C. Huygens (C. Huygens, *Astroscopia compendiaria tubi optici molimine liberata*, 1684).

paper and the main section in finely tooled leather. When telescopes became very long, the tube, usually made of wood, was stiffened by means of stay rods to keep it from sagging under its own weight. In some cases, such as that of Huygens' telescope 123 feet long, there was no tube at all, and the optical parts were aligned by means of a string, held taught by the observer, which joined the eyepiece to the objective **(Fig. 5.3.4)**. The housings of the optical components also became progressively more sophisticated. Often fabricated by lathing high-quality wood — either more common types such as boxwood, or exotic ones such as guaiacum — they were fitted with screw caps to protect the optical parts. (*P.D.S.*)

Catalogue

0.0.1
Anonymous, *Galileo Galilei*, c. 1630
Oil on canvas; 610 x 490 mm
17th century copy; Institute and Museum
of the History of Science (IMSS), inv. 3571-18

The Gioviana Collection in the Uffizi Gallery of Florence consists of 490 portraits of illustrious men from all over the world: heads of state, politicians, military leaders, philosophers, literates, physicians and "scientists". The collection was started in 1552 by the Grand Duke of Tuscany Cosimo I, who commissioned a painter to copy the series of portraits that Paolo Giovio, who had died that same year, had collected in his villa on Lake Como. The collection was enlarged to some extent by the other grand dukes in the Medici family. However, many of the portraits were not copied from the originals, but from frescoes or famous paintings. This is the case of the portrait of Galileo Galilei, which is a rather mediocre copy of one of the other two portraits, much more famous – both found in Florence, one in the Uffizi Gallery, the other in the Pitti Palace – painted by the Flemish artist Justus Sustermans in 1636 and 1640. Galileo is portrayed at the age of over seventy, with many of his discoveries and scientific works behind him, as well as the ecclesiastical condemnation of 1633 that followed the publication of *Dialogo sopra i due massimi sistemi del mondo tolemaico, e copernicano* (1632).

(G.S.)

0.0.2
Egnazio Danti (attr.), *Plane astrolabe*, c. 1570
Brass, wood; 840 x 840 x 860 mm
Replica, IMSS, inv. 3619

This plane astrolabe, of uncertain attribution for many years, was very probably built by the Dominican Egnazio Danti – born Carlo Pellegrino – during the years when he was court cosmographer to Cosimo I. Deriving from a tradition rooted in Hellenistic and Alexandrian times (3rd century B.C.E.-4th century C.E.), the instrument exemplified the essence of mathematical astronomy. It was in fact a refined calculation instrument with a fixed part (the plate), which bore the grid showing the azimuthal coordinates of the place of observation (in this case Florence) and a mobile part (the *rete*), which bore the principal circumferences of the celestial sphere (the Tropic of Cancer, the ecliptic, the celestial equator and the Tropic of Capricorn) and some forty indicators of reference stars. The revolving arm fitted with a sight was used to measure the height of the Sun or that of a reference star above the horizon so as to determine the hour of the day or night, as well as to effect alignments of the instrument and to carry out various operations. Medieval and Renaissance treatises – including the *Trattato dell'uso et della fabbrica dell'astrolabio* (1569) by Danti himself – usually described fifty or so operations that could be carried out with this kind of instrument. When Galileo returned to Florence in 1610, Grand Duke Cosimo II granted him permission to use Danti's plane astrolabe for some of his own astronomical calculations. It is for this reason that the instrument became known as "Galileo's Astrolabe".

(G.S.)

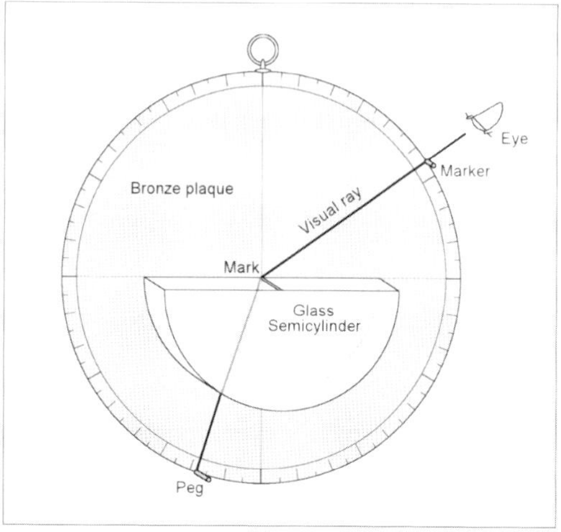

1.1.1
Observation tube
Opera Laboratori Fiorentini (OLF)

In Antiquity, theories on human vision were based on the concept of extromission, the belief that the eye emitted visual rays which reached objects and give a "tactile" perception of them, more or less clear depending on the distance. Among the believers in this concept, the use of the "observation tube", fleetingly mentioned by Aristotle in his *De generatione animalium*, was a common practice. If an empty tube was placed before the eye, the visual rays were not dispersed laterally but were all concentrated to the front, toward the object sighted. It was thought that the tube could facilitate astronomical observations, by allowing the viewer to distinguish the faintest stars. The observation tube was also used to indicate the celestial bodies in teaching astronomy, and as an aiming system for graduated instruments. Representations of the observation tube have come down to us from Greek antiquity (e.g., on the fragment of a vase dating from the 5th-4th century B.C.E. now in the Acropolis Museum in Athens) as well as from the Middle Ages and the Renaissance. In these representations the tube often appears in the shape of a trumpet pointed toward the sky. This has given rise to rash historiographic interpretations attempting to demonstrate that the telescope existed ages before the time of Galileo.

(G.S.)

1.1.2
Ptolemy's experiments on reflection and refraction
OLF

The first known experiments on the phenomena of reflection and refraction, dating from the 2nd century, were conducted by the Alexandrian astronomer and geographer Claudius Ptolemy. In his *Optics* – which has survived only in an incomplete Latin version – Ptolemy examined the behaviour of light rays, both when touching the surface of a mirror (plane, concave or convex) and when traversing the surface that separates two transparent mediums (air and water, air and glass, glass and water). In the first case, his experiments confirmed the law of reflection already known in the past, according to which the angle formed by an incident light ray with the vertical at the point of contact was equal to the angle formed by the reflected ray. In the case of refraction Ptolemy noted instead that, in passing from one transparent medium to another, the light ray underwent deviation dependent on the density of the mediums it contacted. In passing from a rarefied medium to a dense one, the angle of refraction was less than the angle of incidence; in passing from a dense medium to a rarefied one, the angle of refraction was instead greater than the angle of incidence. Utilising a graduated plaque as reference, Ptolemy completed his study on refraction by giving a first definition of the numerical relationship between the two angles in the different cases.

(G.S.)

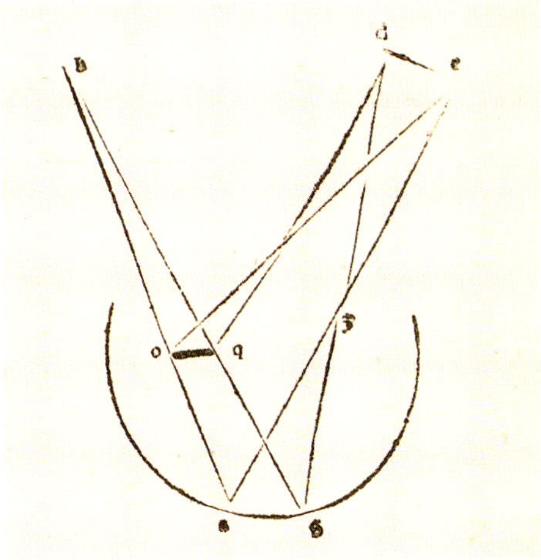

1.1.3
ALHAZEN & WITELO, *Opticae thesaurus ... libri septem*, Basileae, ex Officina episcopiana per Eusebium episcopium & Nicolai f. haeredes, 1572
IMSS Library, MED 2068

The *Opticae thesaurus* was the edition in which students of optics were most likely to encounter the works of Ibn al-Haytham and Witelo from the late 16th century. The edition made these two treatises in the perspectivist tradition of optics not only widely accessible; it also tied them together, because it made the connections between the two works explicit. It also broke up the works in easily digestible parts allowing readers a piece-meal appropriation. The edition was prepared by Frederic Risner, a student of the mathematician and educational reformer, Petrus Ramus. In the years before the publication of *Opticae thesaurus*, Risner and Ramus had travelled together to Nuremberg to visit the workshops of the goldsmiths Wenzel Jamnitzer and Hans Lencker. The frontispiece of *Opticae thesaurus* reflects the editors' interest in instruments. It shows topics typically treated in a work on optics: reflection (a man looking in a mirror), refraction (a man standing in a pool of water with "broken" legs), and phenomena categorised as "meteorological" (a rainbow appearing in the sky). The frontispiece gives, however, a central focus and more space to the legendary story of the destruction of the enemies' ships at Syracuse by means of the burning mirrors of Archimedes.

(*S.D.*)

1.1.4
JOHN PECKHAM, *Perspectiva communis*, [Mediolani], Petrus de Corneno, ? 1482-1483
Central National Library of Florence (BNCF), Incunaboli B.3.30
Facsimile

John Peckham, a Franciscan friar and Archbishop of Canterbury, wrote *Perspectiva communis* in the 1270s. Together with the works of Roger Bacon and Witelo it was one of the standard texts in the perpectivist tradition of optics. The first printed edition of the text published by Petrus de Corneno, is faithful to the manuscript tradition. In 1542 Georg Hartmann, an accomplished instrument-maker, published a version, which differed substantially from the manuscript tradition, and which became the basis of all subsequent editions. The discussion of mirrors and reflection in Peckham's *Perspectiva communis* nicely shows how the study of burning mirrors and the study of mirror images were two separate areas of research. In proposition 50 of book II Peckham discussed the inversion of images in a concave mirror drawing on the pseudo-Euclidean *De speculis*. He identified the point *Z* as the point separating objects that have an inverted image from those that do not. The location of the point would be different for every object, but the collection of points corresponds to the modern focal plane of a concave mirror. While we know this, this was not evident to Peckham himself. He did not identify the point of inversion *Z* with the burning point of the mirror.

(*S.D.*)

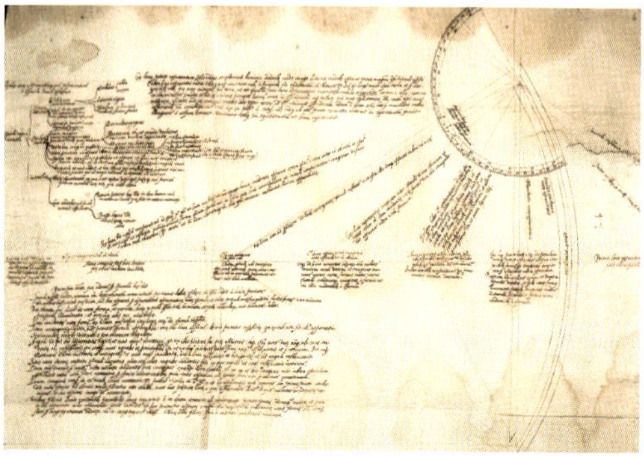

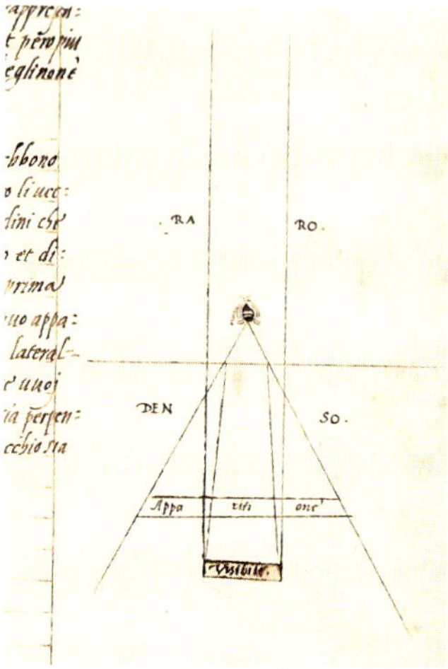

1.2.1
ETTORE AUSONIO, *Theorica speculi concavi sphaerici*, handwritten copy by Galileo, 1592-1601
BNCF, Ms. Gal. 83, f. 4r
Facsimile

The *Theorica speculi concavi sphaerici* is a handwritten copy by Galileo of the original by the Venetian mathematician and physician Ettore Ausonio. In the late 16th century, the original was in the library of Gian Vincenzo Pinelli – one of the most important collectors of books and manuscripts at the time – where Galileo and his friend Paolo Sarpi must have copied it. The original is no longer preserved, but other of Ausonio's manuscripts, related in content to the *Theorica* and at the time also in Pinelli's library, are still in the Biblioteca Ambrosiana in Milan. An adapted version was published by Giovanni Antonio Magini in 1602. The *Theorica* allows the reader to find the locus of an image reflected in a concave spherical mirror given the locus of the object or of the observer. It also located the burning point of the mirror "at a quarter of a diameter" or at half of the radius of curvature of the mirror. Ausonio is innovative in combining the discussion of the burning properties of a concave mirror and the locus of the perceived images in a mirror – two areas of research which had always been kept separate in the perspectivist tradition of optics. The *Theorica* prepared the ground for the introduction of the concept of "*punctum inversionis*" by Giovanni Battista Della Porta.
(*S.D.*)

1.2.2
GIOVANNI FONTANA (attr.), *Della prospettiva*, ms., c. 1460
Biblioteca Riccardiana, Florence, Ricc. 2110, f. 22r
Facsimile

The most likely author of *Della prospettiva* is the mathematician and physician Giovanni Fontana. Perhaps this is the formerly lost treatise on optics which Fontana dedicated to the painter Jacopo Bellini. This manuscript copy was owned by Benedetto Varchi – a member of the Accademia Fiorentina and of Cosimo I's court – before it came into the possession of Giovanni de' Medici who together with Galileo attended Ostilio Ricci's mathematics class. The author of *Della prospettiva* attributed the working of eyeglasses to their magnifying power. He wrote that "the elderly and those who have weak eyesight use eyeglasses, because the likenesses which pass through the glass of the eyeglasses, which is denser than the air, are broken and show the letter larger than it is". Remarkably, the accompanying diagram did not show refraction through curved surfaces as would have been appropriate when explaining convex lenses. The diagram shows instead the magnified image of an object under water.
(*S.D.*)

DE I MIRACOLI
ET MARAVIGLIOSI EFFETTI
dalla natura prodotti.
LIBRI IIII,
DI GIOVANBATTISTA PORTA
Napolitano, nouamente tradotti di Latino in lingua uolgare, & con molta fatica illustrati.
CON DVE TAVOLE, L'VNA DE' CApitoli, l'altra delle cose piu
COL PRIVILEGIO.
IN VENETIA appresso Lodouico Auanzi.
M D L X.

1.2.3
GIOVAN BATTISTA DELLA PORTA, *De i miracoli et maravigliosi effetti dalla natura prodotti, libri IIII*,
In Venetia, appresso Lodovico Avanzi, 1560
IMSS Library, MED 0157

When the polymath, natural magician and theatre-designer Della Porta was in his twenties, he wrote the first edition of *Magiae naturalis*, published in 1558. Numerous editions in Latin and half of a dozen in vernacular languages appeared in the second half of the 16th century. A much expanded version was published in 1589. While the genre to which Della Porta's *Magiae naturalis* belonged is known as "books of secrets" or "experiments", the sections on the burning mirror are mostly collected from written sources. The section on the making of the parabolic burning mirror was taken almost verbatim, including the drawings, from Oronce Fine's *De speculo ustorio* (1551), which also appeared in Italian translation in 1587. Nevertheless, in 1580, upon the request of his patron Cardinal Luigi d'Este, Della Porta also collaborated with Jacomo Contarini, the Superintendent of the Arsenal in Venice, to construct a parabolic burning mirror. Contarini invented a compass which could be used to control the curvature of mirrors. Della Porta's own proposal was borrowed from Fine – including Fine's mistakes.

(S.D.)

1.2.4
ANONYMOUS, *Parabolic burning mirror*, 16th-17th century
Gilded plaster; 240 x 240 x 165 mm
IMSS, Optics M 137

In the 15th and 16th centuries the study of burning mirrors focused on objects of spherical shape, of large size if possible, and presumably designed for warfare. Leonardo da Vinci too conducted studies of this kind, found in the *Codex Arundel* and dating from the decade 1503-1513, relevant to the caustics (envelopes of the reflected solar rays) produced by spherical reflecting surfaces, and the proper shape to be conferred on a concave mirror in order to concentrate all of the reflected rays in a single point. The circulation of a printed Latin translation of the *Conics* by the Greek mathematician Apollonius of Perga – in 1501 only fragmentary by Giorgio Valla, and then limited to the first four books when translated by Giovanni Battista Memmo in 1537, and by Federico Commandino in 1566 – led to technical mastery of the properties of the surfaces generated by rotating the curves obtained by sectioning a cone: circumferences, ellipses, hyperbolas and parabolas. Specifically, it became clear that solar rays could be concentrated in a single point by utilising a parabolic surface. These mirrors, shaped like vases, were positioned with the widest part turned toward the Sun and the narrowest part toward the object to be burned.

(G.S.)

1.2.5
ANONYMOUS, *Concave lens with handle*, 16th century
Rock crystal, wood; 60 x 182 mm
IMSS, inv. 2586

For many years this biconcave lens mounted on a wooden ring with a finely turned handle has been known as the "Lens of Pope Leo X". It was given this name for two reasons: the lens comes from the Medicean collection of scientific instruments, and, in a famous painting by Raphael now in the Uffizi Gallery, Leo X (Giovanni de' Medici) is portrayed holding a very similar lens, used to correct his severe myopia. But the excellent appearance of the lens, totally devoid of the typical flaws found in 16th century glass, has cast doubt on its full authenticity, suggesting that a lens of more recent make (perhaps 18th century) may have been mounted on the older handle. Analysis conducted through the X-ray fluorescence (or more briefly, XRF) technique by the National Institute of Nuclear Physics, Florence and Milan Sections, and by the Department of Electronics and Information of the Milan Polytechnic Institute, has however revealed that the material of which the lens is made is not glass, but rock crystal. This circumstance not only explains the excellent quality of the lens, but also justifies the hypothesis that the object really did belong to Pope Leo X.

(G.S.)

1.2.6
Spheres, mirrors and lenses
OLF

In the 16th century, transparent glass spheres, mirrors and lenses, both concave and convex, were often found in the most important *wunderkammer* (chambers of wonders), that were princely collections of naturalist specimens and scientific instruments. Such objects were especially attractive for their extraordinary property of forming images that were either erect or upside down, enlarged or diminished. Although they had been the object of study for centuries, they still exerted a mysterious, magical fascination on their owners.

(G.S.)

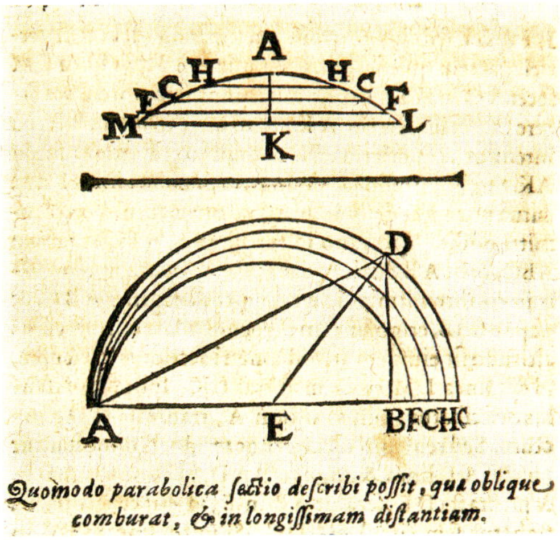

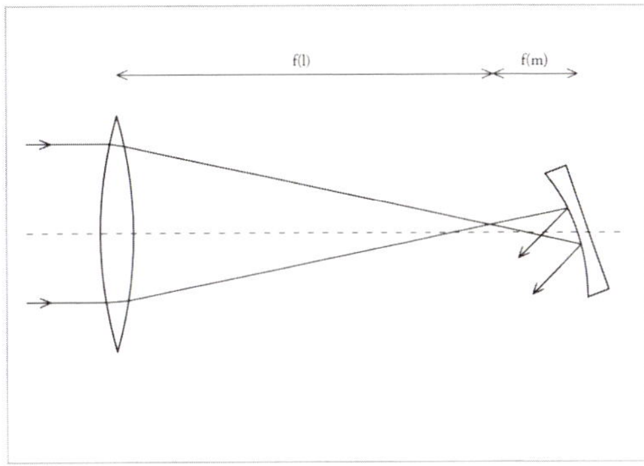

1.3.1
Giovan Battista della Porta, *Magiae naturalis libri viginti, in quibus scientiarum naturalium divitiae et deliciae demonstrantur* ..., Francofurti, excudebat Samuel Hempelius sumptibus Claudij Marnij et haeredum Ioann. Aubrij, 1607
IMSS Library, Med 0378

In his works Della Porta proposed several designs for a *camera obscura* useful for painters or for the projection of images in theatrical productions. In the first edition of *Magiae naturalis* (1558) Della Porta discussed the use of a concave mirror inside a *camera obscura*. In the expanded version of 1589 he proposed to combine a convex lens (in the aperture of the *camera obscura*) with a concave mirror. "Place an eyeglass made from a biconvex lens in front of the hole. From here the image falls on the concave mirror. Place the concave mirror far from the centre so that the images which it receives inverted it will show upright, because of the distance from the centre. In this way, above the hole on the white paper you will see the images of the things that are outside so clearly and openly that you will never cease to be delighted and amazed". This *camera obscura* projected upright and magnified images.
(*S.D.*)

1.3.2
Bourne's telescopic device
OLF

The telescopic device that William Bourne discussed in his letter to Lord Burghley was made of a convex lens and a concave mirror. The light rays went through the lens before they were reflected by the mirror. The observer looked in the mirror with his back towards the lens. The image is upside down and laterally inverted. The total length of the instrument equalled the sum of the focal lengths of the mirror and the lens. Bourne was convinced that the magnification of the instrument was not dependent upon the proportion of the focal lengths of the lens and the mirror, but upon the diameter of the lens. For this reason he insisted that the lens "must bee made very large, of a foote, or 14. to 16. inches broade" (30-40 centimetres), even though the design would have been equally effective with a lens of a smaller diameter. Bourne also pointed out that the convex lens was to be mounted in a frame and "set fast". The observer held the mirror in his hand. Bourne did not mention a tube. The set-up is more reminiscent of a *camera obscura* than of a telescope. Bourne's telescopic combination of a convex lens and a concave mirror is identical to one of Della Porta's *camera obscura* designs.
(*S.D.*)

1.3.3
Model of *camera obscura* with and without lens
OLF

Astronomers had used the *camera obscura* to observe solar eclipses for centuries, but only around 1550 the use of the convex lens and the concave mirror inside the *camera obscura* is documented. The use of the convex lens had the advantage that the image projected by the lens and focused on a piece of paper allowed observers to see more detail than in the image projected through an aperture of the same size to which a convex lens was not applied. The image projected by a concave mirror is no longer left-right reversed (but still top-down inverted) like the image projected in a *camera obscura* without a mirror. For those interested in visual representation the use of the concave mirror inside the *camera obscura* must have been advantageous. When the piece of paper, on which the image projected by a concave mirror was traced, was taken from the wall, and then turned upside down, left and right were as in the original scene. This allowed the draughtsman to easily compare drawing and original.

(*S.D.*)

1.3.4
Anonymous, *Portable camera obscura,* 18th century
Wood, glass, paper; 620 x 205 x 400 mm (open)
IMSS, inv. 3879

The portable *camera obscura* furnished with an objective lens and a flat mirror was long in vogue as a device used by painters who were not exceptionally talented. Some of these instruments consisted basically of a sturdy wooden box fitted with a convex lens inserted in a sliding tube of paper or cardboard. The light received by the lens was projected onto the flat mirror and then onto a glass screen protected by a hood. The image could be put in focus by adjusting the sliding tube, and then traced with a pencil or pen on a sheet of tissue paper placed on the screen. In devices of this type the image appeared erect, but horizontally inverted. Accordingly, after having copied the profile of the image, the artist turned the sheet of tissue paper over and completed the drawing on the opposite side. Since the *camera obscura* was frequently used outdoors, it was exposed to damage of all kinds, which explains why many of the original parts have been replaced with others of more recent make.

(*G.S.*)

L'ARTE
VETRARIA
DISTINTA IN LIBRI SETTE
DEL R. P.
ANTONIO NERI
FIORENTINO
NE'QVALI SI SCOPRINO EFFETTI MARAVIGLIOSI
e s' insegnano Segreti bellissimi del Vetro
nel Fuoco, & altre cose curiose.
Impressione seconda, ricorretta, ed espurgata
da varj errori.
ALL'ILLVSTRISS. SIG. SILVIO ALLI.
IN FIRENZE, Per Marco Rabbuiati. 1661.
Nella Stamperia della Fortuna.
Con licenza de' Superiori.

1.4.1
Antonio Neri, *L'arte vetraria distinta in libri sette ... : ne' quali si scoprino effetti maravigliosi e s'insegnano segreti bellissimi del vetro nel fuoco et altre cose curiose*, In Firenze, per Marco Rabbuiati nella Stamperia della Fortuna, 1661
IMSS Library, Med 0055

On the basis of his experience in glass foundries in Florence, Pisa and Antwerp, Antonio Neri produced the first book on glass-making, in which the process of making glass *à la façon de Venise* was extensively discussed. *L'arte vetraria* consisted of seven books, only the first of which dealt with ordinary and "*cristallo*" glass. The other six books concerned the making of coloured glass. Neri did not discuss how mirrors were made. Until the late 15th century German mirrors were superior. The use of "*cristallo*" glass, invented in Venice, for the production of mirrors was one factor that led to the Venetian excellence in the craft of glass mirror-making. The other factor was that by the mid-16th century Venetian mirror-makers had learned a new technique of applying tin foil to broad glass by means of mercury. This procedure was used to make convex mirrors, but it was impractical for concave mirrors. In the 17th century metal was still the preferred material for making concave mirrors.

(S.D.)

1.4.2
Vannoccio Biringuccio, *De la pirotechnia libri X*, in Venetia, per Venturino Roffinello ad instantia di Curtio Navo et fratelli, 1540
IMSS Library, Geo 167

Although Vannoccio Biringuccio travelled extensively in Germany and Italy, he always returned to his native Siena. In different phases of his life he directed the iron mines in near-by Boccheggiano, before he was appointed to a post in the Sienese Armoury, and later to the function of architect and director of the Opera del Duomo. Biringuccio was thus well-experienced to write the first book in print on metallurgy. The first edition of *Pirotechnia* appeared posthumously in 1540, one year after his death. The book also contained a section on the making of plane and concave mirrors of which "the effects these mirrors produce are miraculous things incomprehensible to me". Biringuccio discussed the making of the moulds, the casting of plane and concave metal mirrors, and the grinding and polishing of the moulds and mirrors on a potter's wheel, first with sand, water and a millstone, and later in the process, "with a ball of wool or of untanned leather tied to a piece of wood, and dressed with calcined tin, tripoli, or ochre". Aware of the limitations of his knowledge Biringuccio concluded that "in this discourse on mirrors, I should be very happy to be aided by some excellent man of intelligence, who would satisfy not only you, but me, in regard to theory as well as practice".

(S.D.)

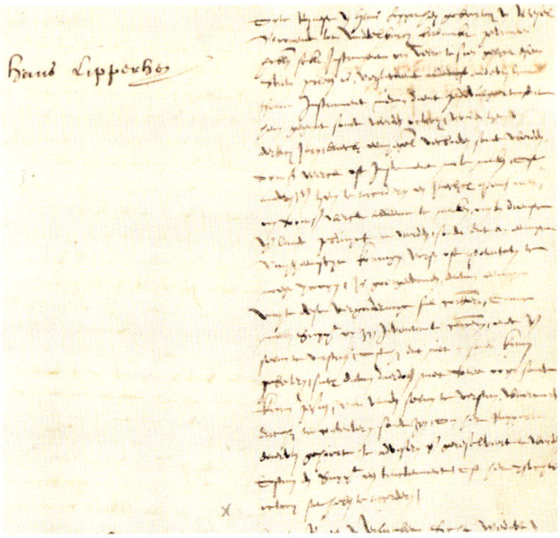

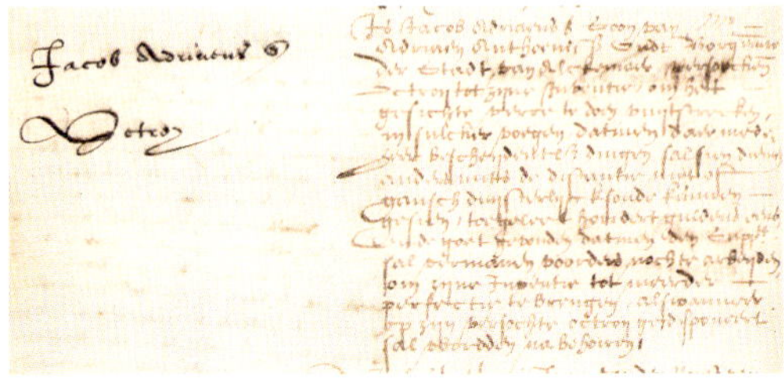

2.1.1
Minute Book of the States General of the Netherlands, Patent Application of Hans Lipperhey, 2 October 1608
Algemeen Rijksarchief, The Hague, Mss. Staten-Generaal, 33, f. 169r
Facsimile

With a letter of recommendation from the government of the province of Zeeland, Hans Lipperhey, a spectacle-maker from Middelburg (the capital of the province) arrived in The Hague (the seat of the national government of the Netherlands). Lipperhey was born in Wesel, in the German region, and had moved to Middelburg in 1594, where he became a spectacle-maker. Middelburg had an important glass-making factory, where the latest Italian glass-making methods were practiced.

In The Hague, Lipperhey first presented his spyglass to Prince Maurice of Nassau, and then turned to the national assembly, the States General, who decided on finances, including privileges and patents. These gentlemen discussed Lipperhey's request either for a thirty-year patent by which others would be forbidden to make the instrument, or a pension under which he would make spyglasses only for the Dutch government. This is the transcript of the meeting of the States General of 2 October 1608, in which Lipperhey's request was discussed. He made several binocular spyglasses for the Dutch government, for which he was richly rewarded. But he did not receive a patent or a pension.

(A.V.H.)

2.1.2
Minute Book of the States General of the Netherlands, Patent Application of Jacob Metius, 17 October 1608
Algemeen Rijksarchief, The Hague, Mss. Staten-Generaal, 33, f. 178v
Facsimile

Jacob Metius (d. 1628) was the son of the well-known surveyor, engineer and mathematician Adriaan Anthonisz called Metius (who served several times as mayor of the city of Alkmaar), and the brother of Adriaan Metius (professor of the mathematical sciences at the University of Franeker). An instrument maker, Jacob kept to himself. When, however, news reached him that Lipperhey had applied for a patent for a spyglass, Metius submitted his own patent petition to the government in The Hague, and this petition was discussed by the States General on 17 October 1608. This is the minute of that discussion.

(A.V.H.)

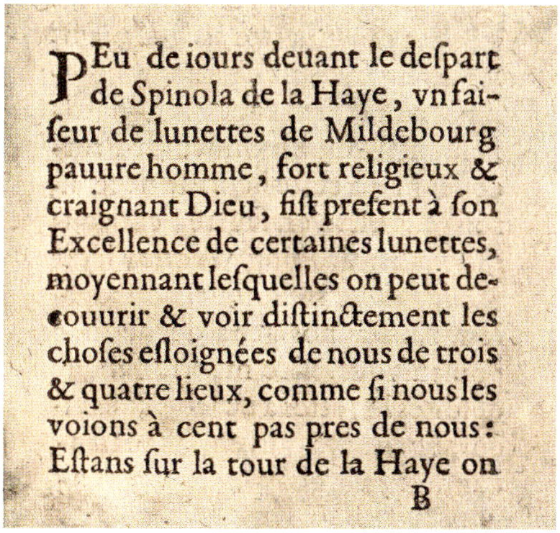

PEu de iours deuant le deſpart de Spinola de la Haye, vn faiſeur de lunettes de Mildebourg pauure homme, fort religieux & craignant Dieu, fiſt preſent à ſon Excellence de certaines lunettes, moyennant leſquelles on peut de-couurir & voir diſtinctement les choſes eſloignées de nous de trois & quatre lieux, comme ſi nous les voions à cent pas pres de nous: Eſtans ſur la tour de la Haye on

B

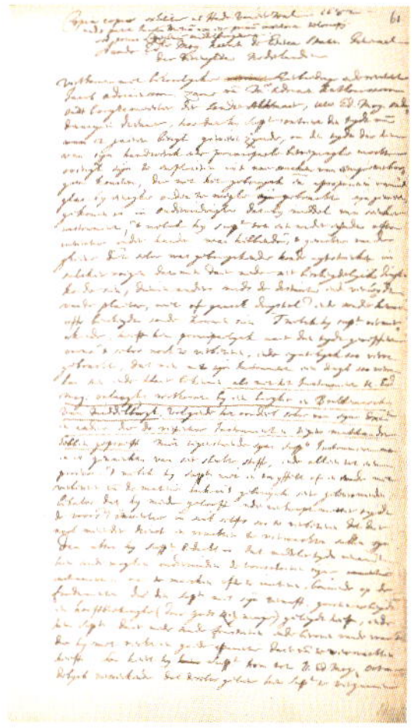

2.1.3

Ambassades du roi de Siam envoyé a l'Excellence du Prince Maurice, arrivé a la Haye le 10 Septemb[re] 1608, [The Hague], 1608

Facsimile

When Lipperhey presented his spyglass to Prince Maurice and the States General in The Hague, at the end of September 1608, that city was the stage for the negotiations for a cease-fire in the war of independence, the Dutch provinces were fighting against the Spanish Crown. Prince Maurice showed the spyglass to Ambrogio Spinola, the Italian commander of the Spanish forces stationed in the southern Netherlands (modern-day Belgium). At the court, there were observers from all countries that had a stake in the outcome of these negotiations, and these observers informed their superiors back home of the appearance of this new device for seeing distant things. The newsletter *Ambassades du roi de Siam* was printed some time in October 1608, within weeks of the events and consisted of two parts. The first part was a report of the arrival in The Hague of a diplomatic mission from the King of Siam. The second part was a description of Lipperhey's presentation of his spyglass to Prince Maurice.

(A.V.H.)

2.1.4

Christiaan Huygens, *Copia copiae exhibitae ab Hadr. Van der Wal 1682, unde patet Jacobum Metium non esse primum inventorem telescopii, sed potius Lippershemium Middelburgensem*, ms.

University Library, Leiden, Ms. HUG 1, ff. 119-120

Facsimile

In the *Dioptrique* of 1637, René Descartes (who was living in the Netherlands at that time) gave the honor of the invention of the telescope to Metius. Half a century later, Christiaan Huygens was putting together the manuscript of his own *Dioptrica* and investigated the various claims. Hadriaan van der Wal, a jurist, showed Huygens a notarized copy of Metius' original petition to the States General, in which Metius clearly mentions that Lipperhey had petitioned the States General before he had. Huygens copied this letter, adding the statement "from which it is clear that Metius is not the first inventor of the telescope, but rather Lipperhey of Middelburg".

(A.V.H.)

2.1.5
Pierre Borel, *De vero telescopii inventore: cum brevi omnium conspiciliorum historia: ubi de eorum confectione* ..., Den Haag, Typographia A. Vlacq, 1655
BNCF, Magl. 15.3.206

Shortly after Lipperhey had applied for his patent, the city father of Middelburg warned the States General that there was someone else in Middelburg who claimed to have the same device and knew how to make it. However, the States General did not identify this person. Almost five decades later, on the urging of Willem Boreel – a native of Middelburg, and then the Dutch ambassador to the French court – the authorities held hearings about the identity of the inventor of the telescope. In addition to the witnesses who came forward to testify in favor of Lipperhey, several others claimed that Sacharias Janssen was the inventor. The testimonies and Boreel's conclusion that Janssen was the inventor were published by Pierre Borel, a physician and Councilor of the future King Louis XIV, in his book *De vero telescopii inventore*. Janssen thus took his place next to the other two claimants for the honor, Lipperhey and Metius.

(*A.V.H.*)

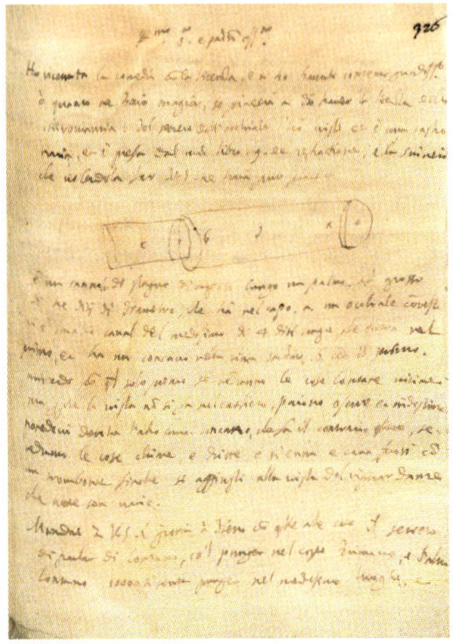

2.1.6
Giovan Battista della Porta, *Letter to Federico Cesi*, 28 August 1609
Accademia dei Lincei Library, Rome, Mss. n. 12 (former Cod. Boncompagni 580), f. 326
Facsimile

Della Porta had investigated the workings of lenses and mirrors for many years, when, some months after Lipperhey's patent application, he heard about the new device for seeing distant things. He was very curious, especially because of the big claims made about the new instrument. Since he himself undoubtedly had at various times combined convex and concave lenses – but lenses presumably not of sufficiently high optical quality – he was disappointed when he finally saw an actual spyglass. In a letter to Prince Federico Cesi in Rome, head of the Accademia dei Lincei of which Della Porta was a member, he expressed his feelings, calling the new device a "*castronaria*" or "hoax", and adding that it had been taken from his 1593 book *De refractione*. However, this book does not contain anything about this combination of lenses.

(*A.V.I.I.*)

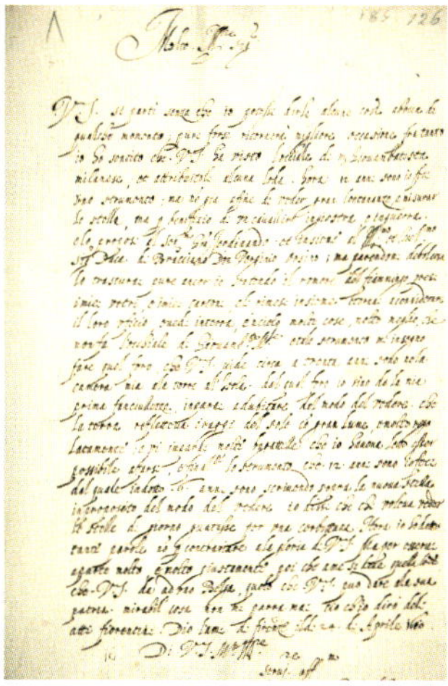

2.1.7
Raffaello Gualterotti, *Letter to Galileo*, 24 April 1610
BNCF, Ms. Gal. 88, f. 126r
Facsimile

Shortly after the publication of *Sidereus nuncius*, Galileo received a letter from the Florentine poet Raffaello Gualterotti, in which Gualterotti told him that 12 years earlier he had made a spyglass for a "cavalry soldier in joust and warfare". But it gave poor results and so Gualterotti placed it aside. This device was presumably put together with ordinary spectacle lenses, which were not ground and polished accurately and therefore produced a blurred image. The invention of the telescope thus was not so much about the particular combination of lenses as it was about the optical quality of those lenses. In Gualterotti's case, the result was "a feeble thing".
(*A.V.H.*)

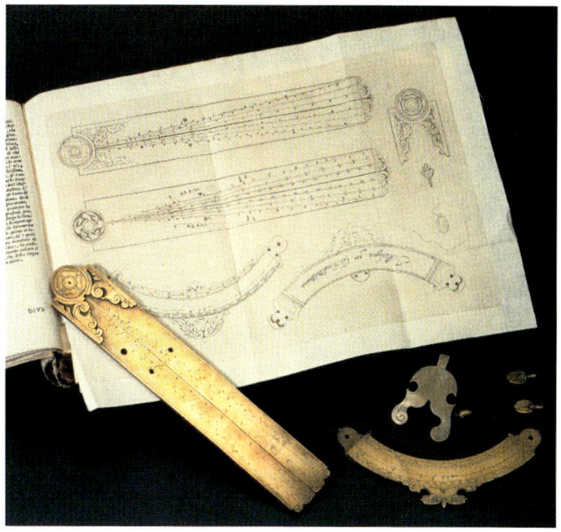

2.2.1
Galileo Galilei, *Geometric and military compass*,
Padua, c. 1606
Brass; 256 x 360 mm (open)
IMSS, inv. 2430

Upon moving to the University of Padua, Galileo decided to annex to his study a laboratory in which to construct scientific instruments. In addition to the ones needed for his own research, of which no trace has remained, he also designed a new type of compass. Starting in 1597, Galileo put on sale examples of this compass, constructed with the aid of the skilled mechanic Marcantonio Mazzoleni. The compass is an elaborate calculation device that can be used to easily carry out many operations in the technical/practical and scientific fields: mathematics, geometry, land surveying, ballistics, and astronomy. The two legs of the compass are engraved on both sides with various graduated scales, and can be combined as required, by means of wing nut bolts, with two accessories: a graduated arc for measuring heights, to be utilised with the aid of a plumb line lowered through a hole at the centre of the hinge in the instrument, and a special kind of bracket (called "*zanca*") to keep the instrument in the vertical position.

Other examples of Galileo's compass are found today in important European and American collections. The compass now in the Institute and Museum of the History of Science in Florence is probably the one donated by Galileo to Grand Duke Cosimo II along with a copy of the treatise *Le operazioni del compasso geometrico et militare* (1606).
(*G.S.*)

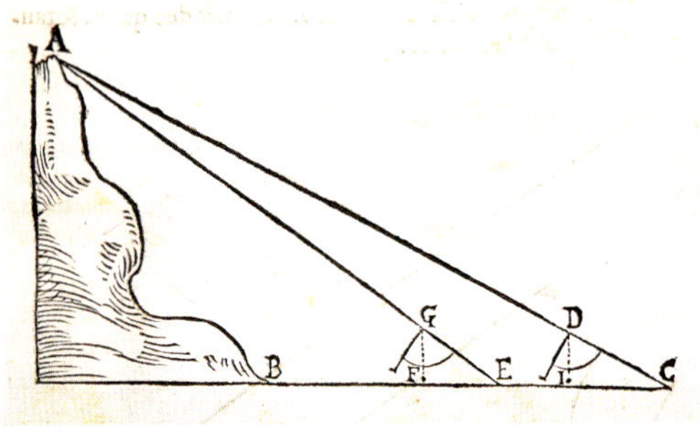

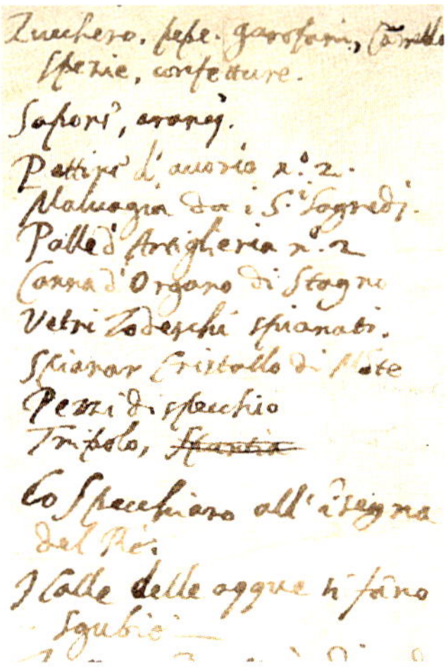

2.2.2
GALILEO GALILEI, *Le operazioni del compasso geometrico et militare*, in Padova, in casa dell'autore, per Pietro Marinelli, 1606
IMSS Library, MED 2023

Although he had been distributing his compass for several years, it was only in 1606 that Galileo published a treatise on it. In addition to the various operations of mathematical calculation, the treatise described how the instrument could be employed to perform measurements useful for land surveyors and geographers, such as determining the height of an inaccessible mountain.

Treatises on fabrication and operation were a widely diffused genre in the scientific literature of the time. While furnishing instructions for using this or that instrument, they also served as their inventors' claim to authorship, in confrontation with plagiarists. That Galileo's instrument was useful and important is proven by the fact that, within a year of the publication of *Le operazioni del compasso geometrico et militare*, the Milanese Baldassarre Capra published a treatise of his own titled *Usus et fabrica circini cuiusdam proportionis* (1607) in which he claimed to have invented the instrument himself. Galileo reacted immediately by publishing a *Difesa contro alle calunnie et imposture di Baldessar Capra* (1607), with which he reduced his adversary to silence. But this negative experience explains why, in early 1610, Galileo did not hesitate a single day to publish his astronomical discoveries.

(*G.S.*)

2.2.3
GALILEO GALILEI, *List jotted down on a letter from Ottavio Brenzoni dated 23 November 1609*, late 1609
BNCF, Ms. Gal. 88, ff. 106r-107v
Facsimile

In the spring of 1609, Galileo began to apply himself to the construction of telescopes. After the first attempts, in which he utilised lenses made for eyeglasses, or at any rate available on the market, Galileo realised that the instrument could be improved only by using lenses with specific optical characteristics. For this reason, toward the end of 1609, he experimented with making lenses, as can be seen from the shopping list he jotted down in the margin of a letter he had just received. Appearing on the list are food items of daily use – lentils, chick peas, rice, raisins, einkorn, sugar, pepper, etc. – but also the basic elements needed to set up an optical laboratory. They range from objects to be used as moulds for concave and convex lenses: "Artillery balls n.° 2 [...] Look for bowls made of iron, or of stone, that is, as artillery balls", to the type of glass to be used: "Flattened German glass"; to abrasives for grinding and polishing: "Tripoli [...] Colophony. Felt, mirror for rubbing", and a memorandum on the shops where he might find the various materials: "The mirror-maker at the sign of the King. In Calle delle Aqque they make chisels". Galileo probably needed these materials to build the most powerful telescope used in his observations of the winter of 1609-1610, whose tube was evidently built of a "Tin organ pipe".

(*G.S.*)

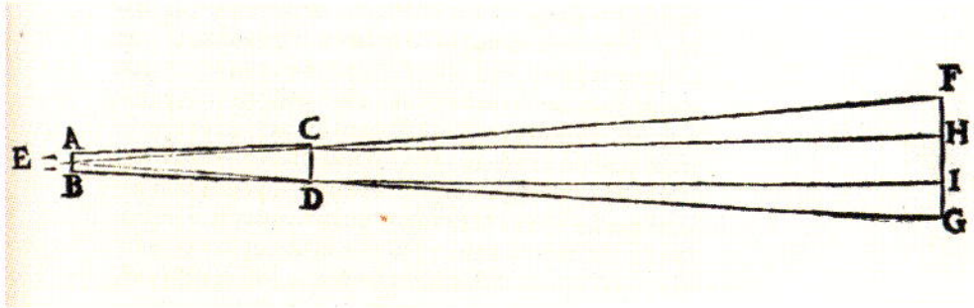

2.2.4
Galileo Galilei, *Sidereus nuncius*,Venetiis, apud Thomas Baglionum, 1610
BNCF, Pal. 1200/23
Facsimile

The *Sidereus nuncius* was written by Galileo very rapidly (in less than two months) with the aim of ensuring the priority of his sensational astronomical discoveries. The proof of this creative urgency, fully justified, is the fact that he travelled from Padua to Venice several times to make addenda to the work while the first parts were already in the stage of typographical composition. The *Sidereus nuncius* opens with a letter of dedication to Grand Duke Cosimo II of Tuscany, dated 4 March 1610, followed immediately by a brief description of the origin, the construction and the upgrading of the telescope. Interestingly, the text states that the thirty-power magnification of the instrument built in 1609-1610 had been improved over the original three-power magnification thanks to a precise knowledge of geometric optics. However the empirical method applied to determine the magnification as well as the very sketchy diagram accompanying the description suggest instead that Galileo must have followed an empirical procedure. Specifically, in the illustrative diagram, the telescope is indicated as a quadrilateral (the section of the tube) marked with the letters *ABCD*, the eye of the observer is at *E*, while the target to determine the instrument's magnification is marked with the letters *FGHI*. No reference is made to the shape of the lenses, and even the path of the light rays from the target to the eye is barely sketched.

(*G.S.*)

2.2.5
Galileo Galilei, *Objective lens*, Padua, late 1609
Glass, 58 mm (410 x 300 mm with frame)
IMSS, inv. 2429

Hoping to establish closer ties with the Medici court and favour his return to Tuscany, Galileo sent a still unbound copy of the *Sidereus nuncius* to Cosimo II along with the telescope he had used for his celestial discoveries in the winter of 1609-1610. As Galileo reported to the Grand Duke in a letter dated 19 March 1610, this telescope was entirely without aesthetic value; and it must have been treated accordingly in the Grand Duke's collection of scientific instruments, considering that it deteriorated almost entirely within a few years' time. The objective lens of this telescope, now separated from the tube, fell to the floor and broke into several pieces a few years before Galileo's death. In 1677, as the myth of the scientist was beginning to take shape, the lens was salvaged, put back together and inserted in a frame made of ebony and ivory, recalling Galileo's discoveries, by the Dutch engraver Vittorio Crosten. Interferometry analysis conducted by the National Institute of Optics has revealed the excellent workmanship of the lens, of the plano-convex type. Analysis conducted through the X-ray fluorescence technique, after a first examination of the data by the experts in the Murano Glass Experimental Station, indicates instead that the surface of the glass has now been damaged by weathering or, in alternative, may have been coated with a thin layer of some kind of "protective" varnish.

(*G.S.*)

2.2.6
Galileo Galilei, *Telescope*, Florence, c. 1610
Glass, wood, leather; 920 x 60 mm
IMSS, inv. 2428

In the same letter of 19 March 1610 with which he sent the *Sidereus nuncius* and his most advanced telescope, Galileo assured Cosimo II that he would find occasion to furnish him with excellent astronomical instruments, outstanding in regards to both optical and aesthetic qualities. After having been appointed "First mathematician" at the University of Pisa and "Primary mathematician and philosopher" to the Grand Duke, Galileo was obliged to keep his promise, as demonstrated by this telescope from the Medicean collection. The instrument was fabricated by a singular construction technique, revealed by the conservative restoration carried out in 1996, which showed that the tube was made of wooden strips. The tube is covered in brown leather, overlaid with narrower bands of red leather. The rich gold-leaf decoration applied by hot-pressing all over the tube and on the two lens housings makes the instrument extremely precious. Only some of the optical parts have survived: the objective is original, while the eyepiece – loosened in the 18th century and entirely lost in the 19th – has been replaced with a biconvex lens dating from the late 19th-early 20th century. At present, the instrument has a magnification of about 20 times.

(G.S.)

2.2.7
Galileo Galilei (attr.), *Telescope*, Padua or Florence, c. 1610
Glass, wood, paper, copper; 1360 x 60 mm
IMSS, inv. 2427

The second of the surviving telescopes of Galilean make shows what the optical instruments used in astronomical observations must have looked like. The tube, consisting of two grooved wooden shells, is held together by copper bands and is covered in paper. The housings of the objective and the eyepiece are also made of wood. The lenses and diaphragms are held in their seats by ring-shaped copper springs. In this case, the optical parts are all original and confer on the instrument a magnification of about 14 with a field of view of only 15 arc-minutes, that is, equivalent to about half the apparent diameter of the Moon. The attribution of the telescope is based on a strip of paper found glued inside the tube specifying the focal length of the objective, in the handwriting of Galileo.

(G.S.)

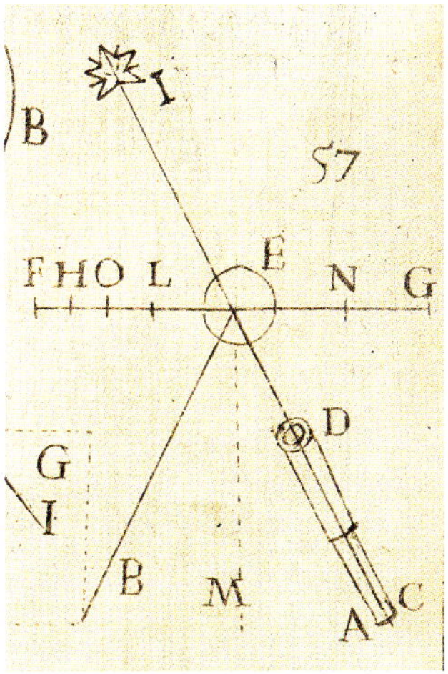

2.2.8
Exploded replica of Galileo's telescope
Jim & Rhoda Morris

Today, external examination of the telescope donated by Galileo to the Grand Duke of Tuscany gives only a vague idea of its beauty, and none at all of its technical complexity. But during the restoration conducted in 1996 it was possible to inspect the internal structure of the instrument and identify its various diverse components. This exploded replica, with and without the leather covering, shows with the utmost clarity the conformation and composition of all of the various parts of Galileo's telescope: from the tube made of wooden strips held together by glue and black cloth, to the lens housings and the diaphragms.
(G.S.)

2.2.9
Giovanni Alfonso Borelli, *Theoricae Mediceorum planetarum ex causis physicis deductae*, Florentiae, ex Typographia S.M.D., 1666
IMSS Library, Med 0773

In Galileo's manuscripts and printed books, no description is given of the device utilised to determine the apparent distances of Jupiter's satellites from the planet with the telescope. A note on the observations conducted on the night of 31 January 1612 states only that the apparent distances were determined with an instrument used to measure the amplitudes of separation. The only reliable description of the device that has come down to us is by Giovanni Alfonso Borelli, a natural philosopher active in the Florentine Accademia del Cimento, founded by the Grand Duke of Tuscany Ferdinando II and his brother Prince Leopoldo de' Medici. Specifically, Borelli dedicated a paragraph in his *Theoricae Mediceorum planetarum* (1666) to a brief description of the so-called "Galilean micrometer", consisting of a ruler divided into 20 equal parts placed perpendicular to the telescope tube, and able to slide back and forth along it. Of this very important device – which transformed the telescope into a precision instrument – Borelli also traced a simple diagram inserted on a plate accompanying the text of the *Theoricae*.
(G.S.)

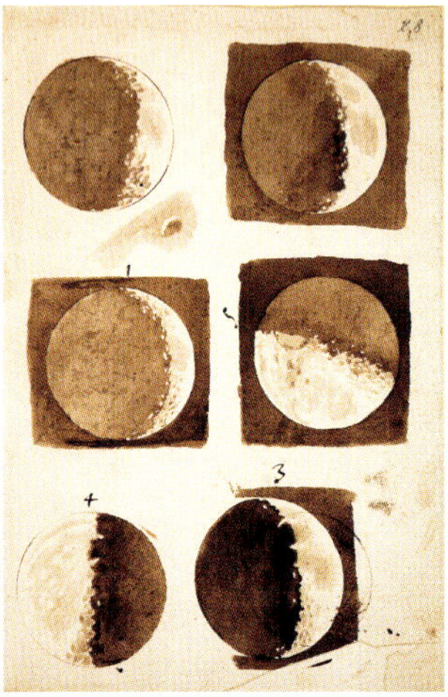

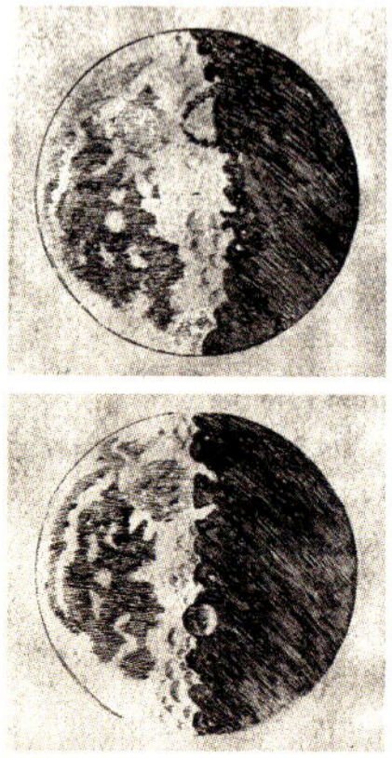

2.3.1
GALILEO GALILEI, *Lunar observations*, ms., 1609
BNCF, Ms. Gal. 48, f. 28
Facsimile

The six autograph watercolours by Galileo (a seventh, marked with the number 8, is found on the *verso* of the preceding page along with the sketch of a horoscope) represent the surface of the Moon during its different phases. These images reveal not only indubitable artistic ability, but above all the great superiority of the telescopes built by Galileo to those used by his contemporaries. Note also, at the upper right on the image marked number 1, the detailed representation of an isolated crater.

(*P.D.S.*)

2.3.2
GALILEO GALILEI, *Sidereus nuncius*, Venetiis, apud Thomas Baglionum, 1610
Facsimile

The *Sidereus nuncius* contains four engravings (one of them repeated twice) of the lunar surface observed through the telescope. These mediocre engravings retain very little of the fascination and realism of the original watercolour plates by Galileo. Two of them represent the Moon in the first and the last quarter. In both, an enormous crater appears in the southern hemisphere, exactly on the terminator, that is, the line of separation between the illuminated part and the part in shadow. Since no formation of such size exists in that region (according to the engraving, it would have had a diameter of about 350 km), it is probably a group of several craters that, due to the play of light and shadow as well as the limited power of the telescope, appeared to Galileo as a single gigantic crater. Historians are not unanimously agreed as to precisely which craters gave rise to this optical phenomenon. Some authors suggest instead that, although Galileo claims in the text to have reproduced the crater as faithfully as possible, he purposely exaggerated its size to emphasise the presence and importance of these lunar formations. It has even been hypothesised that the engraver, in order to show the illuminated side of the crater in an image of such small size (c. 8 cm), had to exaggerate its size; but this suggestion is implausible, considering that Galileo explicitly mentions the vast size of the formation in his text.

(*P.D.S.*)

2.3.3
Teofilo Gallaccini, *Monade Celeste, o vero Trattato di Cosmografia*, ms., 1633-1641
Siena, Biblioteca Comunale degli Intronati,
Ms. L.VI.31, ff. 97v-98r
Facsimile

Galileo's prison sentence, issued on 22 June 1633, was mitigated into a sort of house arrest (*loco carceris*), which he served first at the residence of the Tuscan Ambassador to Rome and then, starting on 9 July, at the Palazzo delle Papesse in Siena, as guest of his friend the Archbishop Ascanio Piccolomini. It was here, in August, on the occasion of "six subsequent observations" conducted from the loggia of the palace "on the same number of evenings", that Teofilo Gallaccini, lecturer in logic and philosophy at the University of Siena, met Galileo. The importance, especially in view of his recent condemnation, of those observations – of which Gallaccini made six sketches, albeit of mediocre quality – lies in the fact that, as we learn from Gallaccini himself, on that occasion all of those present could verify, by means of the telescope, that "the lunar body does not have an even, uniform surface; but one similar to that of the Earth". Despite the fact that Gallaccini had not openly accepted (whether due to conviction or to caution) the revolutionary ideas of Galileo, whose method, however, he enthusiastically appreciated, their acquaintanceship was almost certainly the cause of the fiasco that occurred in the following academic year, when Gallaccini "appeared, but did not lecture because no scholars were present".
(*P.D.S.*)

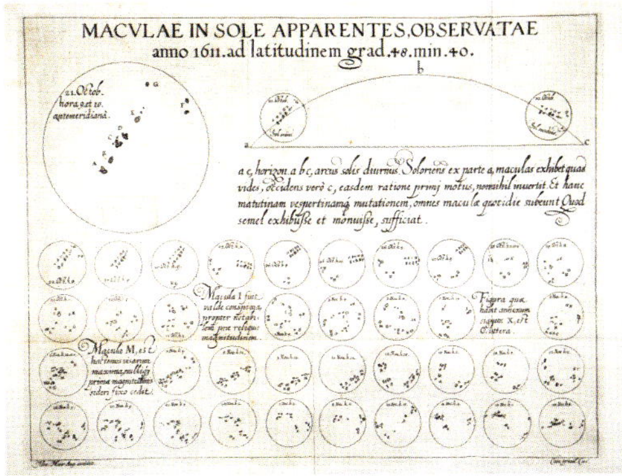

2.3.4
Christoph Scheiner, *Tres epistolae de maculis Solaribus*, Augustae Vindelicorum,
ad insigne pinus, 1612
BNCF, B.R. 178

An advocate of the traditional Aristotelian cosmology, the Jesuit Christoph Scheiner, attempting to salvage the concept of the perfection and immutability of the heavens, sustained that sunspots were caused by clusters of little planets orbiting around the Sun, which, in passing before it, appeared as dark spots. Scheiner set forth his thesis in three letters addressed to the banker Marcus Welser, Duumvir of Augsburg and member of the Accademia dei Lincei. The letters were published in 1612 with the title *Tres epistolae de maculis Solaribus*, under the pseudonym of "Apelles hiding behind his painting".
(*P.D.S.*)

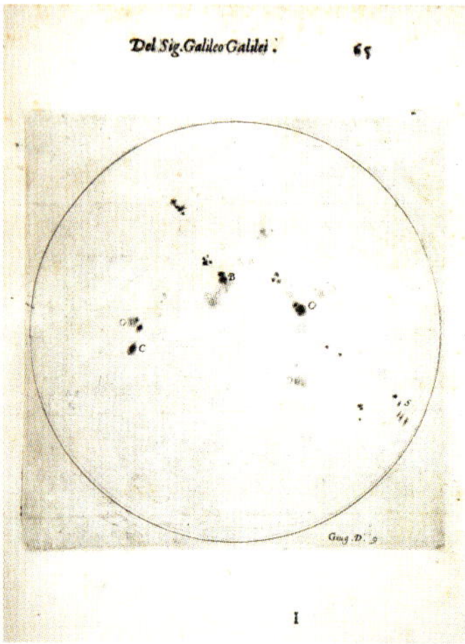

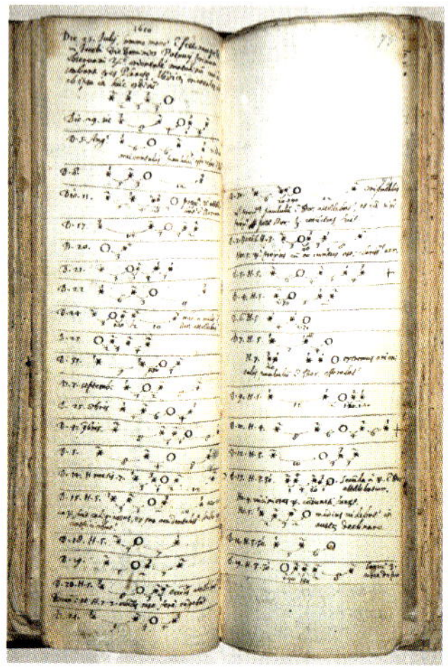

2.3.5
Galileo Galilei, *Istoria e dimostrazioni intorno alle macchie solari e loro accidenti*, In Roma, appresso Giacomo Mascardi, 1613
BNCF, Post. 155

In 1612, Galileo engaged in hot debate with Scheiner over the nature of sunspots and the priority of their discovery. In response to Scheiner's hypothesis, Galileo himself wrote three letters, dated 4 May, 14 August and 1 December 1612, also addressed to Welser. The three letters – in which Galileo presented a long series of observational proofs in favour of his firm conviction that sunspots were contiguous to the solar surface – were published in Rome the following year, at the expense and under the aegis of the Accademia dei Lincei, with the title of *Istoria e dimostrazioni intorno alle macchie solari e loro accidenti*. The work also contains a set of 38 engravings, done by Matthäus Greuter, based on drawings of sunspots made by Galileo using the projection method in June and August of 1612.

(*P.D.S.*)

2.3.6
Galileo Galilei, *Diaries of observations*, ms., 1609-1610
BNCF, Ms. Gal. 49, ff. 73v-74r
Facsimile

Galileo's notebooks on his observations reveal the diligence with which he observed the Medicean Planets in the three years after their discovery. Although he optimistically believed he could rapidly determine their periods of revolution around Jupiter, it took him many months to do so. Due to their brightness, the four satellites are not easily distinguishable from one another, and have extremely short orbital periods (the one closest to Jupiter makes one complete revolution around the planet in approximately 1 day and 18 hours, the furthest away in less than 17 days). This sometimes made it impossible, even when observations could be conducted for two or more consecutive nights thanks to good weather, to identify with certainty the individual satellites. This is probably the reason why the number assigned by Galileo to each satellite is frequently repeated in the course of the same observation. Starting in early 1612 he abandoned this system of notation and began to designate the satellites by vertically aligned dots (four dots for the fourth satellite, three for the third, and so on). Note also that the indications of distance – expressed by a number placed beside an arc-shaped sign – are not all referred to Jupiter, but to the "interstices" between contiguous celestial bodies.

(*P.D.S.*)

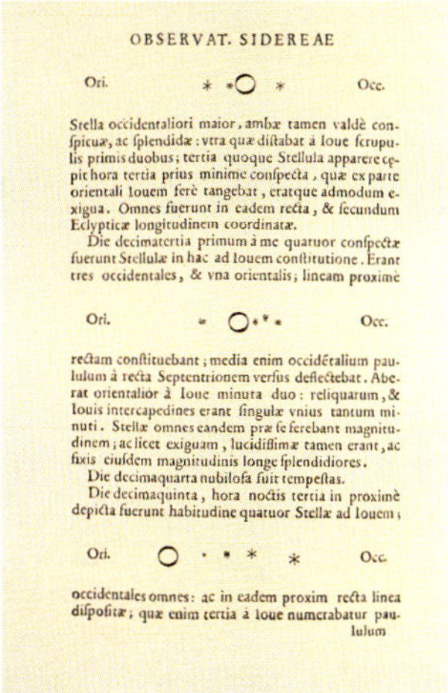

OBSERVAT. SIDEREAE

Ori. * *○ * Occ.

Stella occidentaliori maior, ambæ tamen valdè conſpicuæ, ac ſplendidæ: vtra quæ diſtabat à Ioue ſcrupulis primis duobus; tertia quoque Stellula apparere cępit hora tertia prius minime conſpecta, quæ ex parte orientali Iouem ferè tangebat, eratque admodum exigua. Omnes fuerunt in eadem recta, & ſecundum Eclypticæ longitudinem coordinatæ.

Die decimatertia primum à me quatuor conſpectæ fuerunt Stellulæ in hac ad Iouem conſtitutione. Erant tres occidentales, & vna orientalis; lineam proximè

Ori. * ○*** Occ.

rectam conſtituebant; media enim occidẽtalium paululum à recta Septentrionem verſus deflectebat. Aberat orientalior à Ioue minuta duo: reliquarum, & Iouis intercapedines erant ſingulæ vnius tantum minuti. Stellæ omnes eandem præ ſe ferebant magnitudinem; ac licet exiguam, lucidiſſimæ tamen erant, ac fixis eiuſdem magnitudinis longe ſplendidiores.

Die decimaquarta nubiloſa fuit tempeſtas.

Die decimaquinta, hora noctis tertia in proximè depicta fuerunt habitudine quatuor Stellæ ad Iouem;

Ori. ○ · * * * Occ.

occidentales omnes: ac in eadem proxim recta linea diſpoſitæ; quæ enim tertia à Ioue numerabatur paululum

2.3.7
GALILEO GALILEI, *Sidereus nuncius*, Venetiis, apud Thomas Baglionum, 1610
Facsimile

In the *Sidereus nuncius* Galileo traced the stages that led him to discover Jupiter's satellites, starting from that night of 7 January 1610, in which he saw three stars "small but very bright [...] positioned exactly along a straight line", two of them to the east and one to the west of the planet. When Galileo, without knowing "by what destiny conducted", observed Jupiter the next evening, he found the three stars all to the west of the planet. What attracted Galileo's attention was the fact that in those days, according to the astronomical almanacs, Jupiter should have been moving backward, that is, westward and not – as instead it seemed to have done – eastward. Intrigued by this strange phenomenon, Galileo observed Jupiter again on the 10th (on the night of the 9th the sky was covered), when only two stars appeared to him, both to the east of the planet. Galileo suddenly understood that the changes in position he had observed were not to be attributed to Jupiter, but to the three little stars, which were evidently rotating around the planet. Then on the night of the 13th, he discovered a fourth satellite.

(*P.D.S.*)

2.3.8
ANONYMOUS, *Galileo's "Jovilabe"*, 17th century
Brass; 400 x 195 mm
IMSS, inv. 3178

In this brass version of the jovilabe, two different types of graphic diagram are combined. The instrument was designed by Galileo to determine the position of Jupiter's satellites without having to make complex calculations. To nullify the changing effect of the Earth's rotational movement on the positions observed, their motions could be referred to the Sun. For this purpose, in the lower part of the jovilabe is a pointer that was set at the value of the Earth's position relevant to Jupiter, that is, the Earth-Sun-Jupiter angle, at the moment of observation. This furnished the so-called annual parallax – the Earth-Jupiter-Sun angle – whose value could be read directly on a graduated scale found on the upper edge of the instrument. The pointer was then connected to a rod representing the Earth-Jupiter conjunction, integral with a disc showing Jupiter and the orbits of its four satellites to scale. They are intersected by lines, parallel to the rod, spaced at intervals equal to the planet's radius. In observations, the instrument determined the apparent distance – in units equal to the radius of Jupiter – of a satellite from the planet. The intersection between the line corresponding to this distance and the circle representing the relevant orbit immediately gave the position of the satellite, that is, the Earth-Jupiter-satellite angle, whose value could be read directly on the graduated scale traced on the edge of the disc. When the parallax correction was applied to this value, it gave the position of the satellite relevant to the Sun.

(*P.D.S.*)

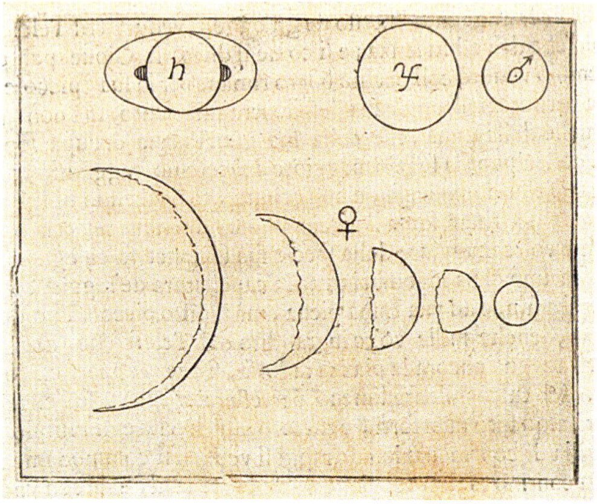

2.3.9
Galileo Galilei, *Il saggiatore*, In Roma, appresso Giacomo Mascardi, 1623
IMSS Library, Rari 153

In 1618 – two years after Galileo's so-called first trial, in which the Copernican concepts were banned – three comets appeared, rekindling the age-old debate over the nature of these phenomena. In March 1619 the Jesuit Orazio Grassi published the *Disputatio astronomica de tribus cometis anni MDCXVIII*, in which he sustained Tycho's theory on comets and, with it, his cosmological system. Prudently concealed behind the pen of his pupil Mario Guiducci, Galileo then composed his *Discorso sulle comete* which was followed by Grassi's reply, published under the pseudonym of Lotario Sarsi, titled *Libra astronomica ac philosophica*. To it Galileo then responded with the *Saggiatore*. Combative starting from the title (the "*saggiatore*" was the precision balance used by goldsmiths to assay gold, an obvious reference to the "*libra*", i.e., the scales, of Grassi), the work goes beyond the controversy over comets to become the manifesto of Galileo's experimental method. One of the engravings illustrating the work shows the planets Saturn (here no longer "three-bodied", but almost reclining on a sort of ring), Jupiter, Mercury and Venus (shown in its various phases), as they appeared through the telescope.
(*P.D.S.*)

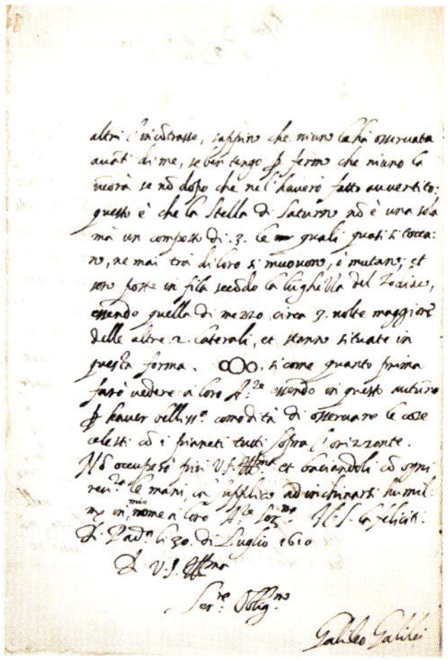

2.3.10
Galileo Galilei, *Letter to Belisario Vinta*, 30 July 1610
BNCF, Ms. Gal. 86, f. 42
Facsimile

While waiting to publicly announce the discovery in a new edition of the *Sidereus nuncius* that he intended to publish, and before breaking the news in the form of a cryptogram to Giuliano de' Medici, Tuscan ambassador to Prague, Galileo announced the "three-bodied" nature of Saturn in a letter, dated 30 July 1610, addressed to Belisario Vinta, Secretary of State of the Grand Duchy of Tuscany. In the letter – aimed at proclaiming to the Medici Court his authorship of the discovery – Galileo included a little sketch of Saturn which, as described in the text, appeared to him to be composed of three different celestial bodies, aligned according to the zodiac and mutually immobile, with the central one three times greater than the two at its sides.
(*P.D.S.*)

3.1.1
Evangelista Torricelli, *Plano-convex objective lens*,
Florence, 1646
Glass, cardboard; 115 mm
IMSS, inv. 2571

Two historic circumstances bear witness to the high technical level achieved by Evangelista Torricelli in the production of lenses. The first is the adaptation of one of his lenses from 1647 to a telescope tube built near the end of the 18th century. The second, more significant, consists of the peculiar investigation carried out at the initiative of the Grand Duke of Tuscany after the sudden early death of Torricelli. The investigation, in which several witnesses were summoned to testify, was aimed at identifying the secret stages of the process through which Torricelli produced lenses that were remarkably smooth and transparent. When carefully examined, the reports of the witnesses do not show procedures very different from those followed by other telescope-makers of the time, Galileo included. Obviously, Torricelli's real secret consisted of exceptionally refined craftsmanship.

Like many lens-makers, Torricelli signed all of his major optical productions (basically, objective lenses), adding to his own name indication of the place, the year and the telescope to which the particular lens was adapted. In the case of this objective, already of considerable size for the time, the signature reads "Vangelista Torricelli. Fiorenza. 1646. Braccia 10 1/4" (equivalent to a focal length of about 6 meters).

(G.S.)

3.1.2
Jacopo Mariani, *Plano-convex objective lens*,
Florence, 1660-1670
Glass; 70 mm
IMSS, inv. 2632

Among those who continued to produce telescopes in Florence were Ippolito Francini, nicknamed "The Thrush", who learned the construction techniques directly from Galileo, and his nephew Jacopo Mariani, nicknamed "The Little Thrush", active after the death of Torricelli. The glass of this objective lens made by Mariani shows a slightly rosy colouration (which may be due in part to alteration of the material caused by exposure to sunlight) and contains numerous bubbles. Traces of the maker's signature, "Jacopo Mari[...] Fior[...]", are still legible along the edge of the lens, amid numerous chips, which may been caused by an attempt to adapt the objective to a smaller tube than the one in which it was originally used.

(G.S.)

3.1.3
EUSTACHIO DIVINI, *Eyepiece with biconcave lens*, Rome, 1665
Glass, cardboard; 55 x 55 x 46 mm
IMSS, inv. 2573

Trained as a clockmaker, Eustachio Divini had become a leading figure in the production of telescopes by around 1649. His production shows a constant tendency toward experimentation, as evidenced by optical combinations that frequently varied from one telescope to another, to the point of using no less than seven different lenses in a single instrument. This eyepiece, signed "Eustachio Divini in Roma 1665 acuto per braccia 4" (that is, for a telescope with an objective having a focal length of approximately 2.3 meters), contains a biconcave lens. This fact shows that, as late as 1665, Divini did not disdain to make instruments whose basic structure was that of the Galilean type telescope.

(*G.S.*)

3.1.4
EUSTACHIO DIVINI, *Eyepiece with biconcave lens*, Rome, 1666
Glass, cardboard; 63 x 63 x 107 mm
IMSS, inv. 2574

This eyepiece too, signed "Eustachio Divini in Roma 1666 acuto per palmi 26" (that is, for a telescope having an objective with focal length of approximately 5.8 meters), is adapted to a basic instrument of the Galilean type. This does not mean that Divini did not also adopt the basic structure of the Keplerian telescope, with a convex ocular lens. Such a structure is in fact found in many of his instruments still surviving today in museums. Divini's pre-eminence in the field of optics had however been severely compromised since 1660 by the appearance on the European scene of two antagonistic figures very different from each other: the Dutch natural philosopher Christiaan Huygens, inventor of a new type of eyepiece with two convex lenses, and Giuseppe Campani, he too originally a successful clockmaker, who turned to the field of optics around 1659.

(*G.S.*)

3.1.5
Giuseppe Campani, *Telescope with eight tubes*,
Rome, c. 1664
Glass, cardboard, leather; 2250 mm (open)
IMSS, inv. 2551

Unlike that of Divini, the optical production of Campani soon became concentrated on a specific model of telescope of the "terrestrial" type; that is, able to furnish erect images. In fact Campani discovered that the basic structure of the Keplerian telescope, having an ample field of view but also known to furnish reversed images, could be improved by inserting between the objective and the eyepiece a combination of two equal convex lenses. The combination produced a second reversal of the images which, added to the first, resulted in erect images. In this telescope, fabricated around 1664, the two lenses are housed in a small cardboard cylinder. It can be inserted in a special tube inside the telescope with one or the other of the two lenses turned inward or outward. Since the lenses are not exactly identical, an inscription specifies that "When this part is inside the tube the object appears larger"; a second inscription states that "When this part is inside the tube the object is seen more clearly".

(*G.S.*)

3.1.6
Interferometry used to verify the shape
of optical surfaces
OLF

When two plates of Plexiglas are held together and observed under the light of a sodium lamp, dark grooves ("fringes") can be seen, along which the reflected light disappears. This phenomenon, called interference, is due to the behaviour of light as a wave, and indicates the gap between the two opposite surfaces (that is, the thickness of the interstice between them). When a finger is pressed against one point or another in the Plexiglas, the thickness of the interspace is reduced and the aspect of the fringes changes. They can be thought of as "stairs", with a height of about 0.0003 millimetres, which as a whole form a map of the gaps between the face-to-face surfaces.

This phenomenon is used in laboratories and optical workshops to verify the shape of a surface to be tested against a sample. For example, to check the convex surface of a lens, a matching concave sample is used. An error in shape appears as a system of sinuous fringes that streak across the face-to-face surfaces, showing where and how much they differ.

(*G.M.*)

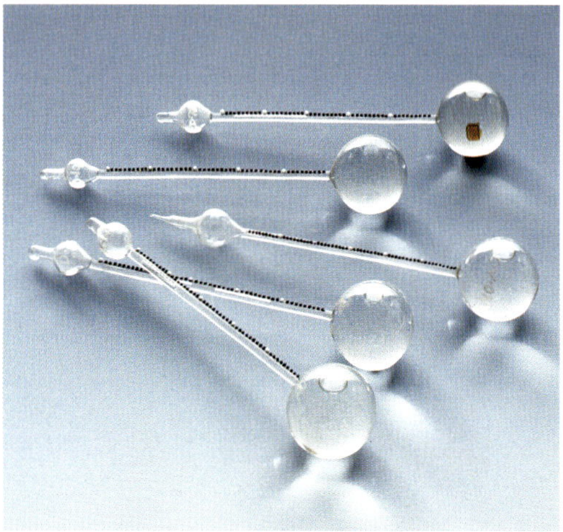

3.1.7
Analysis of colour transmittance in glass
OLF

A glass (a lens) is characterised by the quantity of light transmitted at different colours. In its basic elements, the instrument that measures the so-called "transmittance" consists of a source of light, a dispersing prism, two slits for selecting colours and an electronic detector.

The light emitted by the source contains all of the colours in the rainbow. This light strikes a first slit, which lets pass a thin blade of light. The prism breaks it up into the colours of which it is composed, projecting a rainbow onto the second slit. Through the latter passes a slice of rainbow made of a single colour. The ultra-fine beam of light produced in this way traverses the glass (the lens) to be examined, and is received by the detector, which measures its intensity. By rotating the prism, the rainbow is made to slide laterally into the second slit, allowing a different colour to be extracted each time.

The quantity of light received by the detector is reported on a graph, which shows how much light passes (and how much is absorbed by the glass) for each colour. The structure of this curve, with its peaks and dips, furnishes indications on some substances present in the glass.

(G.M.)

3.2.1
ANONYMOUS, *Fifty-degree thermometer*, Florence, mid-17th century
Glass; 145 mm
IMSS, inv. 39

For a decade, starting from 1657, the members of the Accademia del Cimento conducted an experimental program inspired by the works of Galileo and Torricelli. Many of the instruments used by the members of the Academy in their experiments were made in Florence. The objects made of transparent glass were fabricated by skilled glass-blowers trained in the Venetian glass-making tradition and operating in the glassworks that had been located in the garden of the Grand Ducal residence (Palazzo Pitti) since 1617. Of the over 1,200 extremely fragile instruments made for the members of the Academy (phials, aerometers, balances, thermometers, graduated beakers, etc.), only 120 or so have survived. An interesting fact emerging from the analysis conducted with the XRF fluorescence technique is that the glass of which these instruments is made – for example, the fifty-degree thermometer – contains no lead and does not differ substantially from the glass used in constructing lenses for telescopes. This is still another material proof, albeit a minor one, that in the mid-17th century, no true "optical glass" yet existed.

(G.S.)

4.2.1
Francesco Fontana, *Novae coelestium terrestriumq[ue] rerum observationes* ..., Neapoli, apud Gaffarum, 1646
Facsimile

An indicator of the optical quality of the first telescopes is provided by the observations that were conducted with them, especially those of celestial objects whose details were at the limit of the resolution power of the instruments used. In his *Novae coelestium terrestriumque rerum observations*, the Neapolitan Francesco Fontana stated that he had invented a telescope composed of two convex lenses as early as 1608, discovering only later that it was the same type of telescope presented by Johann Kepler in his *Dioptrice* (1611). In the sequel to his treatise, Fontana displayed beautiful engravings of the lunar surface and some images of the planets, based on observations conducted from 1629 to 1646. While Jupiter, accompanied by the four Galilean satellites, already appears familiar (with two dark bands at the equatorial zone), the engravings of Mars and Venus show that the telescope used by Fontana had some focusing problems. Furthermore, the instrument must have had a resolution power still insufficient to distinguish the ring around Saturn. The body of this planet is sometimes represented in an oblong or sub-quadrangular shape.
(*G.S.*)

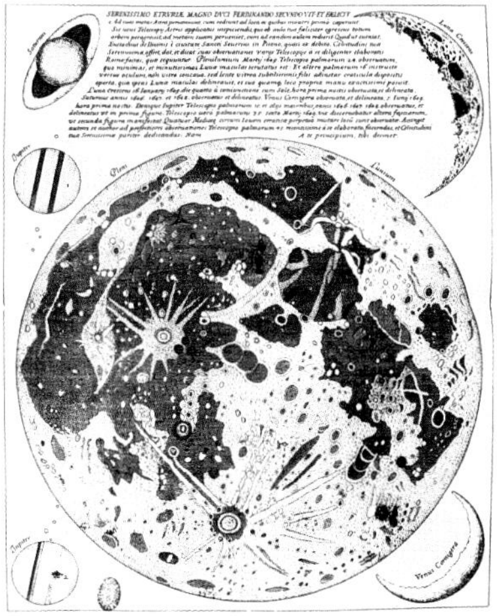

4.2.2
Eustachio Divini, "Serenissimo Etruriae Magno Duci Ferdinando Secundo ...", 1649
Facsimile

Already ranking at the top in the production of telescopes, Divini decided to prepare an advertising leaflet dedicated to the Grand Duke of Tuscany Ferdinando II. The leaflet described the kind of astronomical observations that could be conducted using Divini's telescopes. At the centre of the leaflet was a large image of the lunar surface that was to serve as reference for other mid-century authors. At the corners of the sheet appear Jupiter with the four Galilean satellites, Venus in the form of a thin crescent, and Saturn. The appearance of the latter planet is very similar to the description given by Galileo in the *Saggiatore*, with the central body surrounded by two elliptical loops. But on the whole, the proportions of Divini's Saturn are much closer to reality than those of Galileo's Saturn. Although not yet able to distinguish the ring around the planet, the telescopes built by Divini at this date had a higher resolution power, which was to be improved still further in the following years.
(*G.S.*)

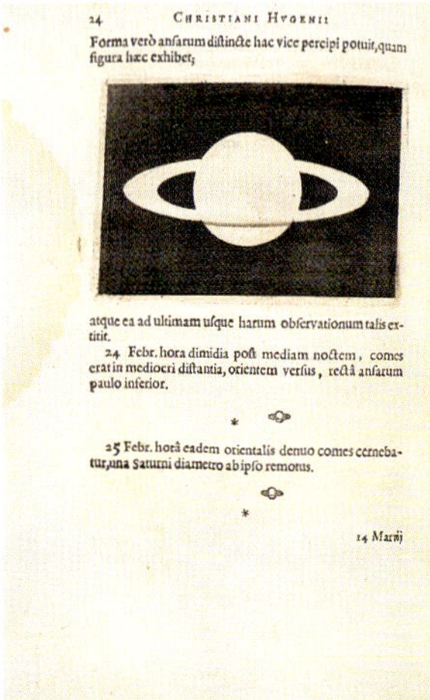

24 CHRISTIANI HUGENII

Forma verò ansarum distinctè hac vice percipi potuit, quam figura hæc exhibet;

atque ea ad ultimam usque harum observationum talis extitit.

24 Febr. hora dimidia post mediam noctem, comes erat in mediocri distantia, orientem versus, rectâ ansarum paulo inferior.

25 Febr. horâ eadem orientalis denuo comes cernebatur, una Saturni diametro ab ipso remotus.

14 Martij

4.2.3

CHRISTIAAN HUYGENS, *Systema Saturnium, sive, De causis mirandorum Saturni phaenomenon, et comite ejus planeta novo*, Hagae-Comitis, ex Typographia Adriani Vlacq, 1659
IMSS Library, MED 1243

Thanks to an eyepiece of his own invention, composed of two convex lenses, Huygens was able to build telescopes that provided clearer images of the celestial bodies. With these telescopes he solved the mystery of Saturn, and also identified its first true satellite (today called Titanus). To explain the enigmatic appearance already noted by Galileo, Huygens hypothesised that Saturn was surrounded by a very thick, solid ring. Being inclined in respect to the planet's orbit, the ring appeared to the observer more or less oblique, depending on the mutual positions of the Earth and Saturn. The inadequacy of the telescopes produced by others prevented a correct understanding of this singular annular structure.

The publication of *Systema Saturnium*, in which Huygens forcefully asserted the superiority of his telescopes, was seen by Divini as a direct attack. Joining forces with the Jesuit Honoré de Fabri, he launched a bitter astronomical controversy. The Divini-Fabri duo sustained that the planet was accompanied by several pairs of large satellites, light and dark, which, assuming various configurations and eclipsing one another, produced the images observed. Divini in particular claimed that it was not the superiority, but rather the inferiority of Huygens' telescopes that made Saturn appear to be surrounded by a ring.

(*G.S.*)

4.2.4

GIUSEPPE CAMPANI, "Eminentissimo Principi Flavio Chisio ...", 1664
Facsimile

Within two years' time, the Divini-Fabri thesis was refuted. A decisive contribution to resolving the controversy with Huygens came from the members of the Accademia del Cimento who, moreover, recognised the ring around the planet even utilising Divini's telescopes. The outcome of this debate was one of the two factors that undermined Divini's reputation, the other being the optical productions of a powerful competitor, Campani, who starting from 1663 did not hesitate to challenge his rival to public comparisons of their telescopes.

Because of their success in these so-called "paragons", Giovanni Domenico Cassini used only telescopes produced by Campani for his astronomical observations. And even when he moved to Paris to direct the observatory built for Louis XIV by the Prime Minister Jean-Baptiste Colbert, Cassini continued to use Campani's telescopes. With them he discovered other satellites of Saturn, distinguished the shadow projected by the ring onto the planet and the shadow of the planet on the ring, and identified the shadows projected onto Jupiter by the four Galilean satellites. Campani exploited these discoveries to promote his own productions, publishing in Rome, in 1664, a pamphlet called *Ragguaglio di due nuove osservazioni* and an advertising leaflet dedicated to Flavio Chigi.

(*G.S.*)

4.3.1
The micrometer and Jupiter's satellites
OLF

The so-called *Galilean micrometer*, used to determine with precision the positions of Jupiter's satellites in respect to the planet, consisted of a ruler divided into equal parts and positioned on the telescope tube at a suitable distance from the eye of the observer, in such a way that each division in the graduated scale coincided with the apparent diameter of the planet. When Jupiter and its satellites are observed through the telescope with one eye, and the micrometer is contemporaneously observed with the other, its image is superimposed on the telescope's field of view, making it possible to measure the distance of each of Jupiter's satellites expressed in radii of the planet.
(*P.D.S.*)

4.3.2
The helioscope and sunspots
OLF

To observe sunspots Galileo utilised a protective method, accurately described in his *Istoria e dimostrazioni intorno alle macchie solari*, invented by his pupil Benedetto Castelli. This model consists of a telescope that projects from its eyepiece an image of the Sun. Using a sheet of paper on which a circle has been drawn, the distance of the eyepiece is adjusted so that the dimensions of the solar disc coincide exactly with those of the circle; the sunspots can then be faithfully traced on the image of the Sun.
(*P.D.S.*)

4.3.3
The satellites of Saturn
OLF

The limited power of his telescope, the reduced inclination of the ring as observed from the Earth in July of 1610, as well as the two dark loops between the inner edge of the ring and the globe of the planet, led Galileo to believe that Saturn was a system of three bodies aligned along the zodiac and mutually immobile. With the model we can verify "experimentally" how an image of Saturn – with its famous ring – when observed through a telescope of suitably low power, can in effect appear exactly as it was described by Galileo.
(*P.D.S.*)

4.3.4
Chromatic aberration and diaphragms
OLF

The negative affects of chromatic aberration can be drastically reduced by decreasing the ratio between the diameter of a telescope's objective lens, the so-called "aperture", and its focal length; an expedient well known to 17th century telescope makers. This ratio can be diminished either by increasing the focal length of the objective while the aperture remains the same or by reducing the aperture while leaving the focal length unchanged. This model, which consists of a telescope aimed at a lighted grid, shows the practical utility of the second method. Due to the relatively large initial aperture, the instrument produces a severely deteriorated image of the grid, where the lines of which it is composed, in reality black in colour, appear surrounded by an iridescent halo. When the aperture is reduced, by placing in front of the objective a diaphragm – that is, a simple opaque screen with a hole in the centre – the image is notably improved.
(*P.D.S.*)

4.3.5
Magnification test
OLF

This model can be used to experiment the method, described by Galileo in the *Sidereus nuncius*, by which "anyone, with little effort" can verify the magnification power of a telescope. A telescope is aimed at a target on which are traced two circles side by side, whose diameters have a ratio of 1 to 20. If the telescope really magnifies 20 times, the smaller circle, when observed through it, will have the same apparent size as the larger circle seen with the naked eye. When both of the circles are observed contemporaneously – the smaller one through the telescope and the larger one with the naked eye – a single circle will appear.
(*P.D.S.*)

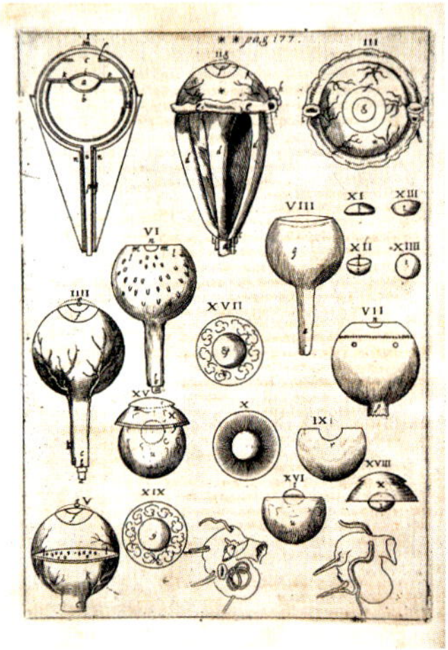

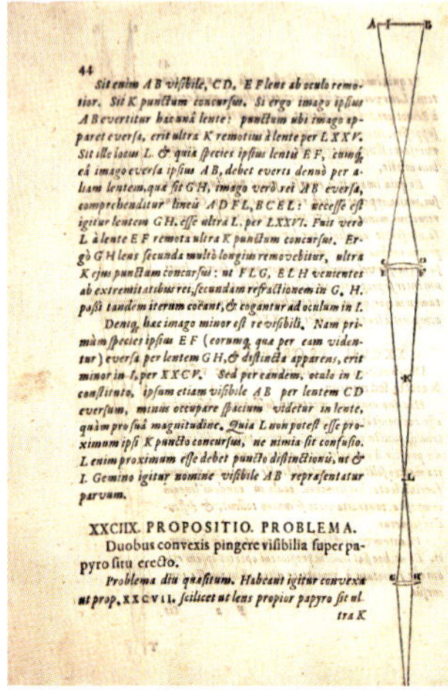

44

Sit enim AB visibile, CD. EF lens ab oculo remotior. Sit K punctum concursus. Si ergo imago ipsius AB evertitur hac una lente: punctum ubi imago apparet eversa, erit ultra K remotius à lente per LXXV. Sit ille locus L. & quia species ipsius lentis EF, eumq; eà imago eversa ipsius AB, debet everti denuò per aliam lentem, quæ sit GH, imago verò rei AB eversa, comprehenditur lineis ADFL, BCEL: necesse est igitur lentem GH. esse ultra L. per LXXVI. Fuit verò L à lente EF remota ultra K punctum concursus. Ergò GH lens secunda multò longius removebitur, ultra K ejus punctum concursus: ut FLG, ELH venientes ab extremitatibus rei, secundam refractionem in G. H. passi tandem iterum coëant, & cogantur ad oculum in I.

Deniq; hac imago minor est re visibili. Nam primùm species ipsius EF (eorumq; quæ per eam videntur) eversa per lentem GH, & distincta apparens, erit minor in I, per XXCV. Sed per eandem, oculo in L constituto, ipsum etiam visibile AB per lentem CD eversum, minus occupare spacium videtur in lente, quàm pro suâ magnitudine. Quia L non potest esse proximum ipsi K puncto concursus, ne nimia sit confusio. L enim proximum esse debet puncto distinctionis, ut & I. Gemino igitur nomine visibile AB repræsentatur parvum.

XXCIIX. PROPOSITIO. PROBLEMA.

Duobus convexis pingere visibilia super papyro situ erecto.

Problema diu quæsitum. Habeant igitur convexa ut prop. XXCVII. scilicet ut lens propior papyro sit ultra K

5.1.1
Johann Kepler, *Ad Vitellionem paralipomena: quibus astronomiae pars optica traditur ...*, Francofurti, apud Claudium Marnium & haeredes Ioannis Aubrii, 1604
BNCF, Magliab. 15.3.211
Facsimile

The term "*optica*" in the title of Kepler's *Paralipomena* is directly linked to the perspectivist tradition of optics based on the treatises of Ibn al-Haytham and Witelo. In effect, in this work Kepler deals essentially with the propagation of light rays in a straight line and the way in which projective geometry can be usefully employed to determine – with appropriate instruments of observation based on the model of the *camera obscura* – the apparent dimensions of the Sun and the Moon, as well as the magnitude of solar and lunar eclipses. The only optical instrument in today's sense (that is, furnished with a lens) discussed in the *Paralipomena* is the human eye, which has, obviously, an essential role in Kepler's pronouncement of a new theory of visual perception.

(*G.S.*)

5.1.2
Johann Kepler, *Dioptrice, seu Demonstratio eorum quae visui et visibilibus propter conspicilla non ita pridem inventa accidunt ...*, Augustae Vindelicorum, typis Davidis Franci, 1611
BNCF, Postillati 25

With the refined mathematical approach distinctive of his works, Kepler also confronts the study of optics in all of its aspects: the reflection and refraction of light rays, and the formation of images in lenses of various shape and size. These studies resulted in the *Dioptrice*, an exhaustive treatise on optics in which Kepler not only examined the structure of the Galilean telescope, but also presented the theoretic diagram of a new type of telescope. Composed of two convex lenses, rather than a convex lens and a concave one, the new telescope soon became known as the Keplerian, or "astronomical" telescope. The significance of this second designation is clear: the Keplerian telescope presented the inconvenience of reversing the images of the objects observed – as is clear from the diagram presented by Kepler himself, in which a sword with the hilt pointed to the left is observed instead with the hilt to the right – and was thus unsuitable for "terrestrial" observations.

(*G.S.*)

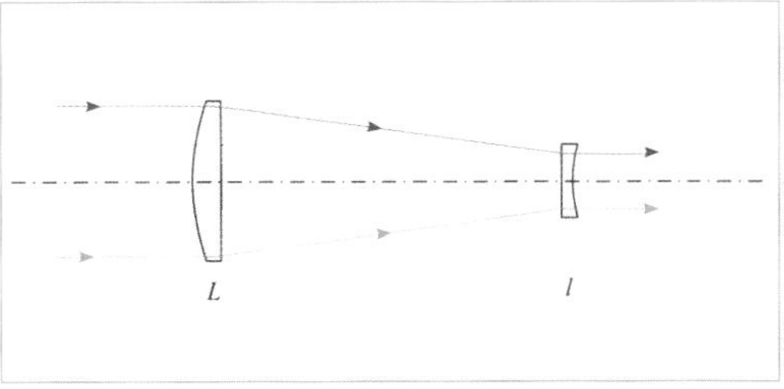

5.1.3
Optical bench with two pairs of revolving lenses
OLF

The structure, as well as the main advantages and disadvantages, of the Galilean telescope and the Keplerian telescope can be clearly understood with the aid of an optical bench with two pairs of revolving lenses. The first pair is composed of a plano-convex lens and a plano-concave lens, the second pair of two biconvex lenses. In addition to the immediate differences in the formation of the images by the individual lenses and their respective combinations – with consequent reduction of the field of view in the Galilean pair, and reversal of the image in the Keplerian pair – the aberrations produced when the optical elements are not perfectly aligned are also apparent.
(G.S.)

5.1.4
Path of light rays in a Galilean telescope
OLF

This model shows the path of the light rays in the optical system of a Galilean type telescope. Two Plexiglas lenses of different section are placed on a plane. The section of the objective lens is semi-convex, that of the ocular lens is semi-concave. A source of light positioned on the optical axis of the lens system projects parallel rays of light. The rays converge toward the focus in front of the objective lens and then become parallel again beyond the ocular lens. If the path of one of the rays is interrupted, the effect of this action on the formation of the image can be seen.
(G.S.)

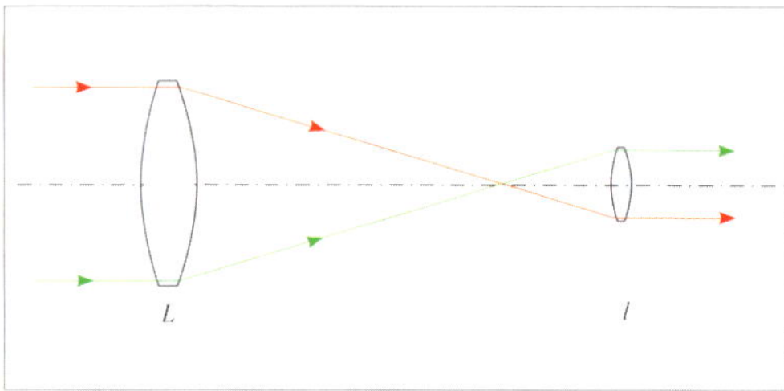

5.1.5
Path of light rays in a Keplerian telescope
OLF

This model, like the previous one, is used to visualise the path of light rays in the optical system of a Keplerian type telescope. The Plexiglas objective lens and ocular lens both have a bi-convex section. The parallel light rays projected by the source converge beyond the objective lens toward the focus. Subsequently, after having been intercepted by the ocular lens, the rays become parallel again. But once the rays have passed through the focal point, their order becomes inverted, showing that the image formed by this optical system is upside-down. In this case too, the effect of interruption in the path of a ray on the formation of the image can be observed.

(*G.S.*)

5.2.1
GIROLAMO SIRTORI, *Telescopium, sive Ars perficiendi novum illud Galilaei visorium instrumentum ad sydera*, Francofurti, typis Pauli Iacobi, impensis Lucae Iennis, 1618
IMSS Library, MED 1013

The Milanese Girolamo Sirtori was one of the first to dedicate a treatise to the construction of the Galilean telescope. After a brief description of the two types of lens that form the heart of the instrument, he explains the method for practical realisation of plane, concave and convex optical surfaces. Sirtori evaluates the materials available on the market – ordinary glass, Venetian "*cristallo*" glass and rock crystal – and then goes on to discuss the instruments to be used for working the selected material: from the reference moulds, to the tools for handling the lenses during the various stages of grinding and polishing, to the abrasives. In his *Telescopium* Sirtori also reports some interesting news on the first diffusion of the telescope, mentioning a spyglass that was delivered to Count Fuentes in Milan by a French gentlemen in May 1609.

(*G.S.*)

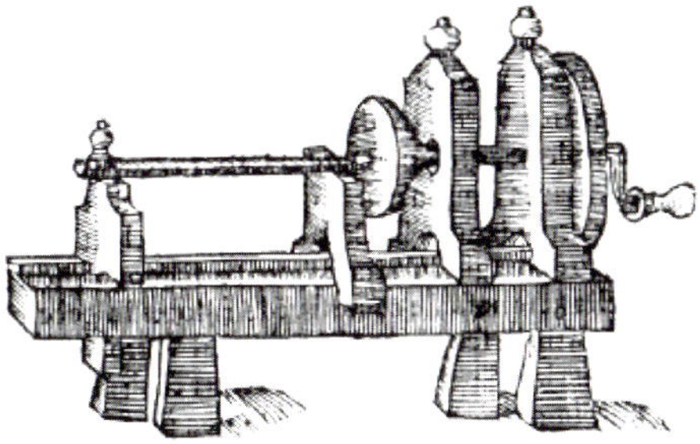

5.2.2

CARLO ANTONIO MANZINI, *L'occhiale all'occhio, dioptrica pratica* ..., Bologna, Herede del Benacci, 1660
Facsimile

Between the time of the telescope's first appearance and the middle of the 17th century, the techniques for grinding and polishing lenses progressed remarkably. *L'occhiale all'occhio* by Count Carlo Antonio Manzini is a treatise dedicated – with no particular theoretical ambitions – to light and refraction, as well as to the various optical instruments that could be fabricated with lenses: from simple eyeglasses to telescopes of the Galilean and Keplerian type (having either one or two lenses) to microscopes. Ample space is dedicated in the treatise to the activity of a new figure in society, the specialised optician, with special reference to the professional experience of Divini, with whom Manzini was in correspondence. *L'occhiale all'occhio* also deals with the selection of glass (ordinary glass, "*cristallo*" glass and rock crystal), with instruments for cutting and grinding the glass, and abrasives used to perform perfect polishing. In addition, Manzini examines some of the machines utilised to make the optician's work easier, faster and more precise. These machines include lathes for preparing the metal moulds and lathes for working the lenses.

(*G.S.*)

5.2.3

JOHANN HOEWEL, *Selenographia, sive, Lunae descriptio* ..., Gedani edita, autoris sumptibus, typis Hünefeldianis, 1647
IMSS Library, MED 2149

Although dedicated to astronomical observations of celestial bodies and, more particularly, to a description of the lunar surface – of which it contains some of the first detailed maps – the *Selenographia* by Johann Hoewel, a brewer by profession and great amateur astronomer, begins with two chapters on constructing telescopes. The description shows the great mechanical competence of Hoewel, who had been working for several years to refine observation techniques by constructing graduated instruments of traditional type and large size, most of them inspired by attentive reading of Tycho Brahe's *Astronomiae instauratae mechanica* (1598). In the case of telescopes, Hoewel's attention was focused on both the optical and the mechanical components, as he now realised that, especially with instruments of large size, the one aspect could not be dealt with independent of the other. One of the engravings in the *Selenographia* offers a particularly interesting view of Hoewel's optical laboratory, dominated by a pedal-driven lathe to grind the lenses using metal moulds of various sizes.

(*G.S.*)

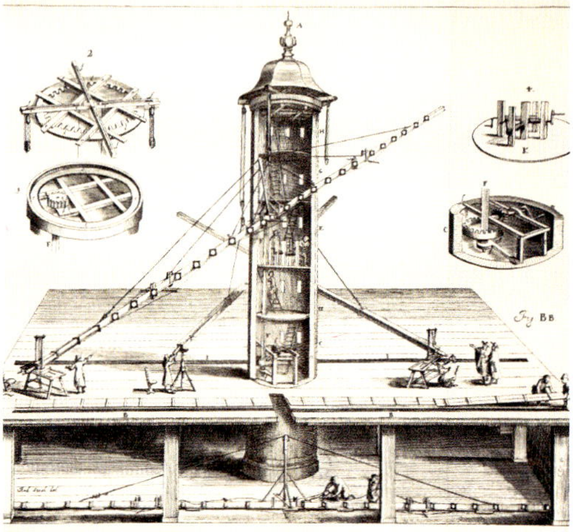

5.3.1
Johann Hoewel, *Machina Coelestis, pars prior*,
Gedani, Auctoris typis et sumptibus, imprimebat
S. Reiniger, 1673
Facsimile

The construction techniques of various astronomical instruments fabricated and utilised by Hoewel are described in detail in this work. The large plate, fig. BB – engraved, like most of the others, by Isaak Saal, on the basis of drawings by the painter Andreas Stech – shows a mount designed by Hoewel, but never built, to be used for big telescopes. The sturdy post used for the mounts of the enormous instruments produced in the second half of the 17th century is replaced here by a big wooden revolving turret, able to house several telescopes contemporaneously. Clearly recognisable in the foreground is the famous 150-foot telescope, with its "à jour" tube (that is, open to lighten its weight) equipped with a complex system of stay rods to keep it from bending. Hanging from the back of the turret (with the eyepiece on the right), the 60-foot telescope. Both were actually fabricated and described in the book. Note the operator inside the turret, intent on actuating the big crank that, through a pair of gears, rotated the entire structure to track the stars. Also visible in the near foreground is the long, narrow, trapdoor that gives access to the level below where the instruments could be stored, protected against the weather. In 1679 the *pars prior* was followed by the *Machina Coelestis pars posterior*. This work, containing a vast collection of data from observations, is extremely rare today because, out of the entire printing, less than one hundred copies survived the fire that devastated Hoewel's observatory in Danzig that same year.

(*P.D.S.*)

5.3.2
Isaac Newton, *Model of reflecting telescope*, 1671-1766
Iron, brass, wood; 300 x 150 x 240 mm
Replica; OLF

The Newtonian telescope is a reflector consisting of a concave parabolic mirror and an elliptical flat mirror. The secondary mirror is inclined 45° degrees from the optical axis and has the function of intercepting the rays of light coming from the primary mirror, deviating them laterally outside the tube, where the eyepiece is positioned. This telescope, which for the simplicity of its fabrication is still today widely used by amateur astronomers, is named after its inventor Isaac Newton, who in January 1668 built one of modest size, with a magnification of around 35, now lost. In the autumn of 1671, Newton built a second example which, on 11 January of the following year, he presented to the Royal Society in London, where it aroused great interest. Among those who examined the instrument were King Charles II, Robert Hook and Christopher Wren, scientist and architect, famous for having designed St. Paul's Cathedral in London and the Observatory at Greenwich. During the course of that same session, Newton was appointed member of the Royal Society. This instrument too has not survived, but parts of a third one built by Newton in the winter of 1671-1672 are believed to have been included in a telescope presented in 1766 to the Royal Society, where it is still found today.

(*P.D.S.*)

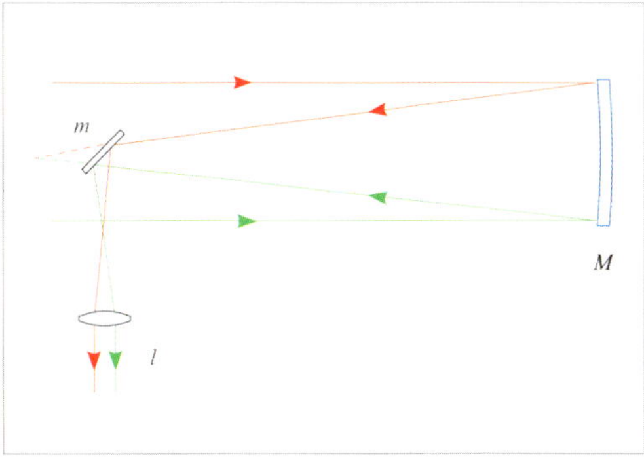

5.3.3
Domenico or Lorenzo Selva, *Gregorian telescope*,
Venice, 18th century
Iron, brass, leather; 400 x 150 x 400 mm
IMSS, inv. 1419

Designed in 1663 by the Scottish mathematician James Gregory, for whom it is named, the Gregorian telescope is a reflector with a concave parabolic primary mirror and a secondary mirror – also concave, but elliptical – that sent the optical beam, through a hole in the primary mirror, back to the eyepiece. Since the secondary mirror is situated beyond the focus of the primary one, the Gregorian telescope furnishes erect images, making it suitable for terrestrial observations as well. This optical combination, no longer in use today, was highly successful in the 18th century, thanks mainly to the famous Scottish optician James Short, who made over 1,360 examples, the largest of which had an aperture, considerable for the time, of 18 inches (c. 46 cm). The telescope displayed here is signed by Domenico Selva, a renowned builder of scientific instruments in Venice, but the attribution is uncertain because his son Lorenzo, who continued his activity, signed all of the instruments with his father's name. The aperture is 65 mm and the tube is covered in leather. The metal rod external to the tube could be rotated to vary the distance of the secondary mirror from the primary one, thus allowing the instrument to be focused.

(*P.D.S.*)

5.3.4
Path of light rays in a Newtonian telescope
OLF

This model shows the path of light rays through the optical system of a Newtonian telescope. Two mirrors of different section are situated on a plane: the objective mirror is concave, the secondary mirror flat. A small section of biconvex lens, placed laterally, represents the telescope's eyepiece. A light source placed on the optical axis of the mirror system projects parallel light rays. All of the rays (except the central one, which is arrested by the flat mirror) converge toward the focus of the objective mirror. They are deviated laterally by the flat mirror and become parallel again beyond the section of the ocular lens. The path of the rays can be interrupted to verify the effect of this action on the formation of the image.

(*G.S.*)

Essential Bibliography

Barkoundah, Y.; Henderson, J., "Plant Ashes from Syria and the Manufacture of Ancient Glass: Ethnographic and Scientific Aspects", *Journal of Glass Studies* 48, (2006), pp. 297-321

Barrera, J.; Velde, B., "A Study of French Medieval Glass Composition", *Archéologie Médiévale* 19, (1989), pp. 81-130

Bedini, S.A., "The Optical Workshop of Giuseppe Campani", *Journal of the History of Medicine and Allied Sciences* 16, (1961), pp. 18-38

Bedini, S.A., "Lens Making for Scientific Instrumentation in the Seventeenth Century", *Applied Optics* 5, (1966), pp. 687-694

Biagioli, M., *Galileo's Instruments of Credit: Telescopes, Images, Secrecy*, Chicago, The University of Chicago Press, 2006

Bucciantini, M., *Galileo e Keplero: Filosofia, Cosmologia e Teologia nell'Età della Controriforma*, Turin, Einaudi, 2003

Camerota, M., *Galileo Galilei e la cultura scientifica nell'età della Controriforma*, Rome, Salerno, 2004

Clagett, M., *Archimedes in the Middle Ages*, Philadelphia, The American Philosophical Society, 1980

Drake, S., *The Unsung Journalist and the Origin of the Telescope*, Los Angeles, Zeitlin and Ver Brugge, 1976

Dupré, S., "Ausonio's Mirrors and Galileo's Lenses: The Telescope and Sixteenth-century Practical Optical Knowledge", *Galilaeana: Journal of Galilean Studies* 2, (2005), pp. 145-180

Dupré, S., "Images in the Air: Optical Games, Magic, and Imagination", in Goettler, C.; Neuber. W. (eds.), *Spirits Unseen: The Representation of Subtle Bodies in Early Modern European Culture*, Leiden, Brill, 2007, pp. 71-92

Dupré, S., "Printing Practical Mathematics: Oronce Fine's 'De speculo ustorio' between Paper and Craft", in Marr, A. (ed.), *The Worlds of Oronce Fine: Mathematics, Instru-*

ments and the Book in Renaissance France, "St. Andrews Studies in Art History", 2, Paul Watkins Publishing, forthcoming

FAVARO, A. (ed.), *Le Opere di Galileo Galilei*, 20 vols., Florence, Barbèra, 1890-1909

GALLUZZI, P., "Evangelista Torricelli: Concezione della matematica e segreto degli occhiali", *Annali dell'Istituto e Museo di Storia della Scienza* 1, (1976), pp. 71-95

GALLUZZI, P. (ed.), *Scienziati a Corte: L'arte della sperimentazione nell'Accademia Galileiana del Cimento (1657-1667)*, Livorno, Sillabe, 2001

GORMAN, M.J., "Art, Optics and History: New Light on the Hockney Thesis", *Leonardo* 36, (2003), pp. 295-301.

GRECO, V.; MOLESINI, G.; QUERCIOLI, F., "Optical tests of Galileo's lenses", *Nature* 358, (1992), p. 101

ILARDI, V., *Renaissance Vision from Spectacles to Telescopes*, Philadelphia, American Philosophical Society, 2007

KING, H.C., *The History of the Telescope*, London, C. Griffin, 1955

KUNICKI-GOLDFINGER, J.; KIERZEK, J., "Central European Crystal Glass of the First Half of the Eighteenth Century", *Annales du 16[e] Congres du AIHV*, London, 2003, pp. 258-262

LINDBERG, D.C., *John Pecham and the science of optics*, Madison, Milwaukee and London, The University of Wisconsin Press, 1970

LINDBERG, D.C., *Theories of Vision from al-Kindi to Kepler*, Chicago, University of Chicago Press, 1976

MINIATI, M. (ed.), *Museo di Storia della Scienza: Catalogo*, Florence, Giunti, 1991

MINIATI, M.; VAN HELDEN, A.; GRECO, V.; MOLESINI, G., "Seventeenth Century Telescope Optics of Torricelli, Divini, and Campani", *Applied Optics* 41, (2002), pp. 644-672

NERI A., *L'arte vetraria*, Firenze, Stamperia de' Giunti, 1612; anastatic reprint edited by BAROVIER MENTASTI, R., Milan, Il Polifilo, 1980

SCHECHNER, S.J., "Between knowing and doing: Mirrors and their imperfections in the Renaissance", *Early Science and Medicine* 10 (2005), pp. 137-162

SIMON, G., *Archéologie de la Vision: L'optique, le corps, la peinture,* Paris, Editions du Seuil, 2003

SMEDLEY, J.W.; JACKSON, C.M., "Medieval and Post-Medieval Glass Technology: Batch Measuring Practices", *Glass Technology* 43, (2002), pp. 22-27

SMITH, A.M., "Getting the big picture in perspectivist optics", *Isis* 72 (1981), pp. 568-589

VAN HELDEN, A., *Catalogue of Early Telescopes,* Florence, Giunti, 1999

VAN HELDEN, A., *Divini and Campani: A Forgotten Chapter in the History of the Accademia del Cimento,* Florence, Giunti, 1981

VAN HELDEN, A., *The Invention of the Telescope,* Philadelphia, The American Pilosophical Society, 1977

VERITÀ, M., "L'invenzione del cristallo muranese: una verifica analitica delle fonti storiche", *Rivista della Stazione Sperimentale del Vetro,* (1985), pp. 17-29

WEDEPHOL, K.H., "Chemical composition of medieval glass from excavations in West Germany", *Glasstechnische Berichte / Glass Science Technology* 70, (1997), pp. 246-255

WILLACH, R., "Der lange Weg zur Erfindung des Fernrohrs", in HAMEL, J.; KEIL, I., (eds.), *Der Meister und die Fernrohre,* Frankfurt am Main, Verlag Harri Deutsch, 2007, pp. 34-126

ZECCHIN, L., *Il Ricettario Darduin, un codice vetrario del Seicento, trascritto e commentato,* Venice, Arsenale, 1986

ZECCHIN, L., *Vetro e vetrai di Murano,* Venice, Arsenale, 1987-1990, 3 vols.